ADVANCING
BIOLOGICAL
FARMING

ADVANCING BIOLOGICAL FARMING

Practicing Mineralized, Balanced Agriculture to Improve Soils & Crops

GARY F. ZIMMER
& LEILANI ZIMMER-DURAND

Acres U.S.A.
Austin, Texas

Advancing Biological Farming

Acres U.S.A.
P.O. Box 91299
Austin, Texas 78709 U.S.A.
(512) 892-4400 • fax (512) 892-4448
info@acresusa.com • www.acresusa.com

Printed in the United States of America

Publisher's Cataloging-in-Publication

Zimmer, Gary F., 1944-
Zimmer-Durand, Leilani, 1971-
Advancing biological farming / Gary F. Zimmer & Leilani Zimmer-Durand. Austin, TX, ACRES U.S.A., 2011
     xii, 244 pp., 28 cm.
     Includes Index
     Includes Bibliography
     ISBN 978-1-60173-019-0 (trade)

1. Agriculture — biological farming. 2. Soil science.
3. Soil plant relationship. 4. Fertilizers and plant nutrition.
I. Zimmer, Gary, 1944- and Zimmer-Durand, Leilani, 1971-
II. Title.

    S605.5.Z562 2011   631.5184

Contents

Don't use Potassium Chloride or other Chlorides
Chloride burns roots ~~dot~~ hooks to available calcium
Calcium —
    The truck of ALL minerals

Mix Humates w/ soil Builders! + Fertilizer
    Potassium —
Do use   Potassium Sulfate - Faster than
    ↑   Potassium Magnesium Sulfate K-Mag
        But check magnesium levels in soil
        And slow release Green Sand (has mag + trace)
                                        Pot
Use sulfate form for Trace Minerals —
    (Not oxides + not chelates)
        homogenized - not pelleted!
        + blend w/ a carbon source like
            humate
    Humates hold onto the minerals so they
    don't leach + make them available

Acknowledgements

Carbon sources - molasses, humates, sugar

I've been on this biological farming journey for over 30 years. Many books, classes, farmers, friends and family have had an influence on how this all came about and developed for me. All the folks at Midwestern Bio-Ag and all the farmers I have worked with all have certainly kept me challenged and informed — telling me what's working and what's not, and letting me know what got them motivated and involved too.

A lot of my hard work and learning has taken place at Otter Creek Organic Farm, where biological farming systems are tested and implemented. Farming together with my family, especially my son Nick and my daughter Sadie, has certainly given me insights on how farmers think and function. Talk is talk, but I feel that walking the talk has given a lot of credibility and confidence to what I say, as well as helping me realize the limitations present.

Developing a way to handle, process and become efficient at local food production and distribution is a major link between biological farmers and consumers. Farmers can do their best to care for the land and produce nutrient-rich, quality foods, but the consumer needs access to those foods, and this is a motivating force for farmers to get involved in that intermediate link.

My family has been very involved in our own farm-to-consumer project — from my wife Rosemarie running the Local Choice Store and providing local quality foods at affordable prices, to my son-in-law Bartlett developing connections between the farm and the consumer and getting more businesses involved in "buying local." Thank you to everyone involved in these projects.

Special thanks to my oldest daughter Leilani and her one-year-old daughter Greta, who forced her to stay home and watch videos of my winter meetings and write this all down. Leilani pored over the winter meeting tapes, asking me questions and making this a "story" for you to read, enjoy, and implement on your farms, in your gardens, and in your food production systems.

Lastly, to change agriculture to a more sustainable, environmentally friendly and high quality food production system requires an educated consumer. It is the support and driving force of these people that is the backbone of biological farming.

Thank you so much.

Gary F. Zimmer

I have had a lot of help and input while working on this book, and I am very grateful to everyone.

In particular I want to thank my husband Bartlett, for spending evenings reading drafts, helping with dinner, and watching the children while I wrote, edited, paced and wrote some more.

I also want to thank Mary Pohlman for tirelessly reading and rereading the drafts, and for all of the positive feedback.

Thank you to my mom Rosie, for all the support and positive feedback, and all of the short-notice help with the children.

Thank you to Ruthann Faber for digging through the boxes of old photographs, looking for pictures of the family farm.

Thank you to Joel Goodlaxson, Lawrence Mayhew, Tim Williams, Bob Yanda, Dan Davidson and Don Faber for contributions, comments and edits.

Thank you to Michelle Broeske for creating the clear, easy-to-understand illustrations.

Writing a book is a huge project, and it can be very overwhelming at times. The help of these people and many others helped me tremendously in moving this project forward and eventually finishing it.

Leilani Zimmer-Durand

About the Authors

Gary Zimmer was raised on a small dairy farm in northeastern Wisconsin. After high school, he left the family farm to attend the University of Wisconsin-Madison where he earned his bachelor's degree in Animal Science. He later moved to Hawaii where he earned his master's degree in dairy nutrition from the University of Hawaii at Manoa. After achieving his educational goals, Gary spent the next five years teaching a course at Winona Area Technical Institute on farm operation and management. As a teacher, he invited outside speakers into his classroom and required his students to conduct on-farm projects so they could apply their knowledge directly to the farm. He left his teaching job to work as a consultant for Brookside Labs, where he was introduced to the concept of mineralizing the soils based on a soil test. He also began reading and studying more about farming, which led him to discover *The Albrecht Papers*, Don Schriefer's books, and many other books on the concept of biological farming, all of which inspired him to test the ideas of biological farming on his own farm.

In 1979, Gary and his family moved to Spring Green, Wisconsin, where, in the early 1980s, he and three partners started Midwestern Bio-Ag. Midwestern Bio-Ag has grown from a small soils consulting company in the Midwest to a much larger consulting, fertilizer and dairy nutrition business with close to 100 consultants and 6,000 farms around the country. In addition to being president of Midwestern Bio-Ag, for the last 15 years Gary has also managed the Bio-Ag Learning Center, a livestock and crop demonstration farm where he tests products and methods for successful biological farming.

Together with his son, Nicholas, Gary owns and operates Otter Creek Organic Farm, a 1,200-acre, 200-cow dairy farm. The herd is fed mineralized corn, beans and forages all grown on-farm, and rotationally grazed on pastures stocked with lush, mineralized forage plants. Otter Creek also raises grass-fed beef, pastured hogs, and free-range chickens, as well as producing its own award-winning line of seasonal raw milk cheddar cheeses. In 2008, the Midwest Organic and Sustainable Education Service named Otter Creek Organic Farm "Organic Farm of the Year" for their commitment to diversified agriculture, environmental sustainability, and community outreach.

For the last 35 years, Gary has been visiting farms around the United States and the world. He has seen many successful biological farms, which has given him a lot of new ideas to test on his own farm. Gary is a highly sought-after speaker, and spends his winter months traveling around the United States and the world speaking to farmers about what he's learned from years of reading, study, observation, and experience as a practicing biological farmer.

Gary is the author of the book, *The Biological Farmer* and numerous articles about biological farming and sustainable agriculture.

Leilani Zimmer-Durand grew up on her family's farm in southwestern Wisconsin. She was exposed to the idea of biological farming at a young age working with her father on the farm as he studied and developed his biological farming techniques.

Leilani received her bachelor's degree in philosophy from the University of Wisconsin-Madison, and soon after moved to Hawaii with her husband where she entered the ecology program at the University of Hawaii at Manoa. Leilani did her master's research on the comparative physiology of native and invasive tree ferns in the Hawaiian rainforest, using a systems approach to examining why some species succeed at the expense of others. After Leilani earned her master's degree, she worked for several years in conservation, field biology, and natural resources management in Hawaii.

After the birth of her son, she moved to Wisconsin and began working for Midwestern Bio-Ag as head of the research and education department. Leilani brings her training in ecology to agriculture, with a focus on a more holistic, systems view to research versus reductionistic science. As such, she is able to bridge the world of university agricultural research and the biological system of farming. Leilani is the author of numerous articles about biological farming and sustainable agriculture.

Introduction

If you have ever been to one of my winter meetings or presentations you know that I talk fast and go over things quickly. This book is a chance for my words to be slowed down and for you the reader to capture what I've been saying, and absorb the ideas at your own pace.

It's been ten years since my first book, *The Biological Farmer* came out. Since that time, I've been traveling around the world, learning, observing, and expanding my knowledge and experience. Also, since the first book our Otter Creek farming enterprise has expanded to have more acres, more crops, more cows and a local food market. It has truly become a passion of mine to produce local, quality foods at affordable prices. Food that is good for the community both in health and in local economic wealth – absolutely essential if we are to be sustainable for our people, our water and our land.

Since the first book, I've now seen and have come to better understand the importance of soil carbon, biology and delivering nutrients to the plant. My understanding of the "system" has expanded to include the concepts of carbon-based fertilizers, homogenized blends, and balanced quality nutrients that are root friendly, pH controlled, and hooked to carbon so that leaching and erosion are reduced. I believe that adapting this system worldwide would have a huge beneficial effect on our environment and on the efficiency of food production.

So where's the research, you might ask? I'm sure you can find some pieces somewhere in the world, but this isn't about studying bits and pieces; it's about the "big picture." It is about pulling it all together into a coherent, workable, effective system.

I've often joked that an appropriate title for this book would be "Since the Book" because of all my added experience over the past ten years. We considered this title, but the book in your hands is really all about a plan and a farming system. It's about mineralizing the soil and keeping nutrients, plants and soil life in balance.

This book is my story; a story gleaned from my winter meetings from 2003 through 2005, written and edited by my daughter Leilani.

You see, I've been telling the same story for more than 20 years now, each time with a slightly different approach in the telling of it, but always with the same message. The facts about soils don't change much over time. It's our understanding and management of them that does. We are at a point in time where a huge farming change is now possible, and I don't mean better biotechnology, chemicals, farm equipment or seed genetics. There are upcoming opportunities to produce healthier, better crops relying less on modern "improvements" and more on the things we've learned by improving fertility in a natural, sustainable way for many years. We at Midwestern Bio-Ag and the many farmers involved in biological farming keep on learning, experiencing, observing and making improvements in production methods every year.

So please, when you read this book don't be too quick to judge. Don't read between the lines. I'm sure you can find some details you won't or can't agree with, but remember, these are my thoughts, observations, ideas and experiences up to this point in time. Show me a better way and I'm ready to make changes and take on new ideas after they have been tested and their success demonstrated on the farm. I want to know when it works, how it works, why it works or doesn't work. If a new idea makes sense, improves quality and/or yield, and is profitable, then let's go with it!

It seems so simple — to know where you are going and to have that plan — and it can be. Get the basics right first. Don't keep looking for a "silver bullet" fix. Use common sense. Study science. Watch reality. Question everything.

My purpose in writing this book is to open your mind to a new perspective on farming. If I can't change your way of thinking, I can't change your actions. If I don't change how you think, how you perceive soil life and its role in successful farming, how you think about food and how food is grown, and how you perceive the concept of "health," then I can't change where you are going.

I hope you'll enjoy reading this book and that it will encourage you to get involved whether you are a farmer, consultant, student, educator or consumer.

Gary F. Zimmer
April 30, 2010

Riding my favorite horse, Taffy, with my daughter Leilani in 1975.

Chapter 1

How I Got Started in Biological Farming

I got my start in farming growing up on a small dairy farm in Hortonville, Wisconsin. Because I was young I wasn't really paying attention to what we were doing, I just followed what my father did. Back then, there really was only one way of farming. The terms "biological," "sustainable," "conventional," and "organic" to describe food and farming didn't exist. Farming was farming, and you did the best you could to get a good crop and feed your family.

To my father, farming was a battle. Each year we battled weeds, we battled insects, and we battled diseases. We didn't have the tools or the knowledge that modern-day farmers have to combat these problems. Instead, year after year we fought an exhausting battle against these threats with the tools we had, which mainly consisted of our own two hands.

Before the introduction of DDT and other insecticides, herbicides and fungicides, crops were organic because we didn't put anything synthetic on our crops or soils. We weren't doing what we could to make the system work biologically — we didn't understand the role of soil life or the need to remineralize the soil — but we also weren't adding any chemicals that would damage soil life.

When DDT came along, it seemed like a miracle. My father thought it was a fantastic innovation that was going to make farming so much easier, almost like he'd died and gone to heaven. It took away a part of the battle, but he didn't understand the costs. Back then I don't think anyone did.

It was a different era, when everyone trusted doctors, politicians, and the man selling farming chemicals. When that salesman showed up at the door and told us he had a new chemical called Atrazine that would kill weeds and was so safe you could drink it, we didn't ask questions we just poured it on. And we weren't alone. The theme of the era seemed to be newer is better; more is better; too much is not enough. Early in the era of chemicals, farmers were starting to double their crop yields, and were seeing less damage from diseases and insects. It made poor farmers look good. When an insect or disease did come along, sweeping through a population of plants or animals, the entire crop wasn't lost. The new chemicals were increasing yields and decreasing the amount of work needed to farm, so of course they were widely adopted.

One memory really sticks out for me from the DDT era that highlights how trusting we were. Every harvest season, a threshing crew came to our farm and we prepared big dinners for them. My mom and sisters made these beautiful pies and this wonderful, homemade food from the fruits, vegetables and meat grown right on our farm. Before the threshing crews came in for lunch, someone went through the dining room with a sprayer full of DDT and squirted it right on top of the food to kill the flies. No one even considered there could be a problem with this. I didn't either at the time.

As I got older I started to question this new system of farming. I decided not to stay on and take over my father's farm — I was in my early 20s when I left the farm and went to college. I studied Animal Science at the University of Wisconsin, and later earned my masters degree in Dairy Nutrition from the University of Hawaii.

After college, I got my first job working at a technical college in Minnesota, teaching farm operation and management. It was a small program — there were only two instructors — so I had to teach agronomy, soil science and farm finance. My background was limited to my education in dairy nutrition and my experience growing up on a farm, so I brought in a lot of outside speakers to present on topics I wasn't comfortable teaching. It gave my students and me a chance to evaluate what people from different sides of the farming world were saying, and to come to our own conclusions on whether or not their message made sense. It really opened my mind to a lot of different ideas about farming.

Each summer I gave my students small research projects. One of the students came up with the idea of putting calcium and sulfur on hay fields. At the end of the summer he took feed tests, and we were surprised to see that the hay had changed. I had never before considered that what you put on the field shows up in the plants. If you put on a lot of manure, you have more nitrogen (N) and potassium (K) in your hay. If you put on calcium (Ca) and sulfur (S), you have more calcium and sulfur in your hay.

This little experiment really got me thinking. As a dairy nutritionist, I started to wonder: what would ideal hay look like? And what would we need to put on the land to get that ideal hay?

A couple years later I took a job with a soil consulting company in Minnesota. The company's philosophy was to balance your soils based on a soil test, which was a new concept to me at the time. I visited a lot of farms where I took soil samples and then went over the soil test with the farmer and made fertilizer recommendations. I did have some success with this approach, but I also had some failures. I still didn't see the soil as a living system, and I had a lot to learn about fertilizer sources and application. One of my biggest difficulties came down to justifying why I was making certain fertilizer recommendations, and then finding a good source of those fertilizers at a decent price for the farmer.

The consulting company I worked for spent a lot of time training consultants on how to take soil tests and make recommendations. Maybe it was me, and maybe it was confusion within the company, but I never did get a clear understanding of why to recommend one type of fertilizer over another. Within the company there was disagreement about what types of fertilizer to recommend to farmers. Some of the older consultants advocated using all naturally mined minerals, like rock phosphate and potassium sulfate. They strongly believed these were better sources of nutrients, but they never explained why, at least not in a way that I understood. The younger consultants, on the other hand, believed you could go ahead and use DAP (diammonium phosphate), potassium chloride and other salt fertilizers and it wouldn't matter as long as you applied enough calcium and didn't use anhydrous ammonia. There was a lot of debate about these two approaches at our annual meetings, but I never heard a clear explanation of why one approach was better than the other.

I later learned that what was missing from this discussion was an understanding of the solubility of chemical fertilizers compared to mined minerals, and the effect this had on soil life and plant performance. It took me many years of reading and studying to figure that out for myself. At the time, I always recommended naturally mined minerals because intuitively it seemed like the better choice. In the end it didn't matter what I recommended because there wasn't an affordable source of the products — in those days they just weren't available.

I'll never forget the first crop farmer I worked with who had no animals. I took his soil samples and saw that his phosphorus (P) levels were very, very low, about 10 percent of where I liked to see them. Because he had no animals he had no manure, so I recommended he apply soft rock phosphate. It wasn't available locally, which meant it was expensive and it could only be purchased by the semi-truckload. The problem was that he didn't want 24 tons of rock phosphate. And when he asked me why I recommended rock phosphate rather than another source, I couldn't really explain to him why rock phosphate was a better soil corrective than DAP or MAP (monoammonium phosphate). In the end, he took my recommendation for phosphorus and went up to the local co-op and bought what they had available.

I ran into this same situation more than once in my years working as a soil consultant. Even though there were a lot of educational resources within the company, none of them answered my questions about why one fertilizer was better than another, why using anhydrous ammonia was problematic, which calcium sources were better for which situations, or even why we always recommended calcium. I was frustrated that I didn't have any good answers for the farmer to justify my recommendation of soft rock phosphate over DAP, or potassium sulfate over potassium chloride. I eventually left that company to see if I could find answers to my questions elsewhere.

After I quit soil consulting, my wife and I bought a small farm in southwestern Wisconsin. I was eager to see what I could learn by applying what I knew about soils to my own land. I was teaching agriculture at the local high school, running my own small farm, and at the same time I was gathering pieces of the biological farming puzzle. I spent a lot of time gathering information on soils and farming, reading every book I could and trying to make sense of it all.

Through all of this reading and studying I was beginning to understand the importance of balancing minerals in the soil, and of using soil tests as a way to address limiting factors. I was also learning the difference between soil correctives and crop fertilizers and the role each plays in producing a healthy crop. I read a lot of different authors: from William Albrecht and Rudolf Steiner to Louis Bromfield and Don Schriefer. Don Schriefer's books, *From the Soil Up* and *Agriculture in Transition* were a big influence on me. They focus on the importance of managing soil air, water and decay, and explore basic principles that I still follow today. It was an important piece of the puzzle, but I still had questions. What about exchangeability of minerals? When we put something on the soil are we sure the plant can get it? And what about mineral interactions? We want to apply sulfur to help pull off extra magnesium (Mg), but does sulfur then interfere with selenium (Se) or copper (Cu)? There were a lot of questions out there for which I still had no answers.

Meanwhile, I was trying to farm biologically based on what I'd read. I wasn't using any herbicides, and I had a two-row corn planter, a two-row cultivator, and no money. I struggled away. Our farm was in a narrow valley in the hills of southwestern Wisconsin, and I also rented some ridge land. One year I decided to plant corn on the rented land, and because I didn't have much money I bought some urea and borrowed a buggy to spread it. It was a brand new buggy, not a scratch on it, and I hooked it behind my truck to drive it up the steep, winding access road to get to my rented fields. About halfway up that steep, bumpy road the pin hitching the buggy to my truck came out. I felt a jolt up in the cab of the truck and turned around in time to see that brand new buggy careening backwards down the hill behind me. My heart just about fell out the bottom of my stomach. I stopped the truck and jumped out to chase the buggy and see where it landed. It went rolling backwards into the woods along the side of the road and jolted to a stop. I thought I was sunk. When I got to the buggy, I saw the wheels had gotten wedged between two trees, and there was not a scratch on it. Someone was looking out for me that day! On the other hand, maybe it was a sign that I wasn't supposed to be spreading urea in the first place. At any rate, I backed my truck down the hill to the buggy, rehitched it, pulled it out from between those two trees, and continued on my way.

My educational background is in animal science and dairy nutrition, but in order to understand biological farming I knew I had to learn more about soils. A big breakthrough for me in understanding biological farming was seeing the parallels between how a cow's rumen functions and how soils function. Both are biological systems that require microorganisms to function efficiently, and both are a "stomach" — the rumen provides nutrients to the cow, while the soil provides nutrients to plants.

Similarities between the Soil and a Stomach

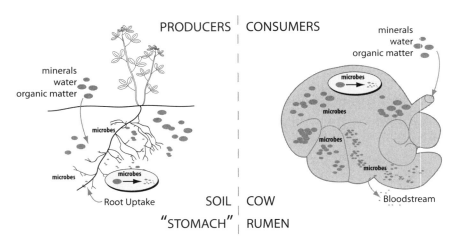

The rumen is essentially a fermentation vat, and in order to function well the cow's stomach requires minerals, water and organic matter in the form of feed. Microorganisms in the cow's stomach make the feed digestible, and without them the cow could not absorb nutrients. Because it is a biological system, you have to balance what goes into it. You can't violate the principles of a cow — feed too much grain and the cow can get sick. To keep the cow healthy and performing well you need to feed it a balance of fiber, protein, starch and minerals.

It is a similar situation with plants and soils. The soil is essentially the stomach of the plant. It's where the plant gets its nutrients. Like the rumen, the soil requires inputs like minerals, water and organic matter in the form of residues, manure or compost. And similar to the rumen, the soil cannot deliver nutrients to growing plants very efficiently without the help of microorganisms. In a natural system, in order for nutrients to be converted into a form plant roots can absorb, bacteria, fungi and other soil life need to be present. An exception to this is a farming system where soluble nutrients are applied in a form that is readily plant available. However, if you add too many soluble nutrients, it is similar to feeding a cow too much grain. It upsets the balance in the soil, and the soil and plants become stressed. The soil needs to be fed a balance of minerals and enough organic matter (OM) to keep the soil life functioning and nutrients available for growing plants.

Soon, more and more pieces of the biological farming puzzle started coming together for me. Putting together this puzzle in my mind is similar to how my oldest daughter described learning a foreign language. My daughter Leilani spent her senior year of high school in Denmark as a foreign exchange student. She lived with a Danish family that spoke limited English and she attended a Danish high school where all of her subjects were taught in Danish. Of course, all of the Danish high school students could speak English, but to keep up with the class my daughter needed to be able to understand what was said in Danish. By sitting in class she could pick up some new words, but it wasn't easy. Just imagine sitting all day in school listening to someone talk in Danish. If they speak as fast as I do, how do you sort out what they're saying? It's got to be really challenging and really boring.

School started in August and Leilani was picking up new Danish words every day. But just increasing her vocabulary didn't mean she could really understand the context and meaning of what was being taught. She could pick out more words here and there all the time, but putting it all together and understanding complex discussions wasn't possible. In October, her host family took her on a holiday to a small Danish island. All this time my daughter had been increasing her Danish vocabulary, and while on holiday, speaking the language all day with her host family, she had a breakthrough. One night she went to bed and had a dream in Danish. It was like a light switch flipped in her mind. When she woke up the next morning she found that instead of being able to understand words here and there in conversations, suddenly she could understand the conversation. There were still words missing, but instead of picking up on pieces and not understanding the whole, she could understand the whole and would just miss some of the pieces. And as time went on, she could understand more and more of those pieces and put them together.

I had a similar experience learning about biological farming. For years I read and studied and picked up on small pieces like the importance of not using harsh fertilizers that would harm soil life or the importance of having lots of earthworms to keep channels open for air and water in the soil. But I didn't understand how things fit together: how the biological farming puzzle pieces meshed with one another. Until one day when it all just seemed to click. There was no one event, no one book, no one piece of information that tied it all together — it just suddenly came together in my mind. There were still bits and pieces here and there that I didn't understand, but I now had a good grasp of the concept of biological farming.

Soon after this breakthrough I decided to start soil consulting again. I put an advertisement offering soil consulting services in a small magazine called *Natural Farming* and I got a few calls, but progress was slow. Then one day I got a call from a man named Fred Kurschner. He and several partners were starting a small company in southern Wisconsin that sold compost turners and biological stimulants. We met for lunch one day, and after talking for several hours about biological farming, he offered me a partnership in their business as a soil consultant. The head of the company was Ralph Engelken, a farmer from Iowa who had written a book titled, *The Art of Natural Farming and Gardening*. They needed a soil consultant to make the business grow and soon I started working with them as an equal partner in the business.

I quit my job, then teaching high school agriculture so I could devote all of my time to soil consulting and farming. That summer I was asked to speak at my first field day, held at Ralph Engelken's farm in Iowa. I took my 8-year-old daughter, Sadie with me to the field day and she sat down in the middle of the front row to hear me speak. It was a hot July day and everyone was uncomfortable. I had a flipchart along with just two graphics: a triangle with "physical, biological, chemical" at the corners, and a list of bullet points on the benefits of using calcium. I was used to speaking about farming from my years of teaching high school, and I thought it was going pretty well. Then, three quarters of the way through my speech, Sadie stood up and walked out of the tent. She was in the front row, so everyone sat there and watched my daughter walk out on me. She was the first person ever to walk out when I was speaking — I was shocked! Well, it turned out there was an ice cream truck around the corner and she decided not to wait until I finished to get herself some ice cream!

The following December, Ralph Engelken died unexpectedly. The remaining company partners met to discuss what to do. Only three of us decided to stay on: Fred Kurschner, Don Gilbertson and me. We brought in my brother-in-law, Donald Faber as our financial officer and changed the name of the company to Midwestern Bio-Ag.

I started developing fertilizer blends and came up with a 10-9-10-13S corn blend with trace minerals, and a 5-8-12 alfalfa fertilizer with calcium, sulfur, magnesium and trace minerals. I also developed a calcium product with soluble calcium in it called Bio-Cal. Those three products were very successful and all remain staples of the company to this day.

I visited a lot of farms that first spring and summer and sold enough fertilizer to keep the business going, but when fertilizer sales stopped for the winter we ran out of money and started wondering what to do next. We had to keep the business going through the winter if we were ever going to survive. We looked at a lot of different options, and the one my partners liked best was that I would utilize my skills as a dairy nutritionist to consult on dairy farms. I'll admit that I really did not like the idea. I had done dairy nutrition work before and each farm I worked with was like taking on a wife and 40 kids. But we forged ahead with the idea and began selling kelp along with mineral blends I developed for heifers, dry cows and lactating cows. This line of products was also very successful and is still being sold by Midwestern Bio-Ag today.

So we managed to survive through our first year and soon after I started writing booklets on biological farming. My first one was called "Soil-Stomach-Silo" and made the comparison between soil health and animal health. A couple years later I came up with my "Six Rules of Biological Farming." I believed then, and still believe today, that if we are going to be a successful biological farming business we need to educate farmers on the principles of biological farming.

In 1991 Midwestern Bio-Ag purchased a demonstration farm where we could test new products and new farming ideas so we could share our success and failures with others each year at our field day. In 1994 my son Nicholas and I purchased the dairy farm across the road, which we transitioned to organic production. I now spend each spring and summer working on the farms, and each fall and winter traveling and teaching about biological farming.

I've been in agriculture my whole life, and I have seen a lot of changes come about. To me, the move towards larger farms and more chemical-intensive agriculture over the last 80 years is "the great forgetting." I think that as we moved toward higher production and the use of chemistry for easy fixes, we forgot what it means to work with nature and with the soils.

Even though tractors made farming easier, my father always said a lot of the fun went out of farming when he stopped farming with horses. Tractors were faster and more efficient, but they were also noisy and too far from the ground. My father loved the quiet of sitting on a plow behind a team of horses, listening to the harnesses creak and the blackbirds sing on the fence posts. He missed the warm earthy smell that came from freshly turned ground. He missed just sitting and watching the roots and earthworms as he worked his soil. The tractor, with its speed, its height, and its noise took all of that away. And I think that when we removed ourselves from the ground, we forgot our connection with the soil.

My belief is that whether we work with dairy or with soils, we can't violate the principles of this earth. We need to work with soil, plant and animal biological systems in order to produce healthy crops and livestock without an over-dependence on chemicals and fast fixes. We need to return to a more natural system of farming, a biological system, in order to make farming sustainable for the long haul.

Chapter 2

What The Farmer Can Do

Good farm management is doing something about something you can do something about. Farmers can't do anything about the weather, but they can grow bigger root systems that are more drought tolerant. Farmers can't do anything about the sandy soils, or the heavy clay soils they have, but they can grow cover crops, add manure and compost, apply calcium and sulfur, use balanced fertilizers, and practice thoughtful tillage to promote soil life and healthier soils. Farmers cannot do anything about the length of the growing season, but they can interseed cover crops or plant late season cover crops, shorten their crop rotation, or add another type of crop to the rotation to increase plant diversity. And there are many things farmers can do something about: tight, compacted soils with poor water infiltration; bare soils over winter and poor soil life; minerals in the soil and the health of the crops they grow.

As a farmer, I have got jobs to do. I need to control air and water. I need to manage minerals. I need to manage plants and microbes. I won't maximize my yield by adding more chemical fertilizers and using more pesticides — in my opinion we have gone as far as we can down that road. Better yields require better management. Trace minerals, calcium, big root systems, healthy plants, loose, crumbly, live soils, proper placement of nutrients at the proper time — these are the things that increase yields. Management is more than just N, P, K and pH. Or, to look at this from

What Does It Take to Grow a Good Crop? more than just fertilizer!

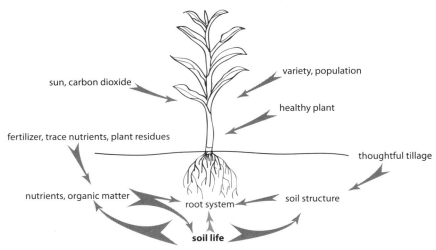

sun, carbon dioxide

variety, population

healthy plant

fertilizer, trace nutrients, plant residues

thoughtful tillage

nutrients, organic matter

root system

soil structure

soil life

another angle, if I get my soils loose and crumbly with lots of life in them and an abundance of balanced minerals, what is there to limit my yields?

Farm management practices have changed over the years, based on the knowledge and tools available to farmers. It's interesting to take a look back and see how we got to the place we are at now.

1930s: Plough, Dig and Disk. The 1930s was the decade of the Great Depression. Farm prices and farm income were at an all-time low. In the 1930s most farms had no electricity and relied on real live horse power rather than tractors. Farming was a battle, and the farmer's main tool was the plough. Farmers didn't understand how to manage tillage and control erosion, and that eventually led to the Dust Bowl when all the heavy tillage farmers were doing simply eroded away the topsoil. Farmers didn't understand the importance of soil life and the damage caused by too much tillage like we do today.

1950s: Kill, Crush and Control. The post World War II boom in America brought a lot of changes to agriculture. More and more farmers were using tractors rather than horses and mules to farm. Chemicals like anhydrous ammonia, DDT and herbicides came on the market, and were treated like a miracle that made farming easier. Suddenly, instead of fighting to control weeds and insects, a farmer could spray on a chemical and the problem was gone. In the beginning these "magic bullets" worked fairly well, but no one understood the negatives and the long-term effects.

1970s: Spread, Strangle and Spray. The 1970s brought the introduction of no-till farming, and also the first organic faming organizations. Across the nation farms were getting larger, and more and more chemicals were used each year. The new Farm Bill encouraged fencerow-to-fencerow farming. New genetics were introduced that produced higher-yielding plants. Even though there was some awareness of the downsides of chemical application, most people and most farmers still didn't question their use. The 1970s were boom years for agriculture, before the farming recession of the early 1980s hit.

2000s: Patent, Police and Pollute. The turn of the century brought widespread use of GM (genetically modified) crops, patented by companies who closely policed their use. As highlighted by an infamous lawsuit against a canola farmer in Canada, any farmer found to have genetically modified genes in their crop and no record of purchasing GM seed could be sued by the seed company for patent infringement. In addition, more and more farmers were practicing no-till, yet runoff from excess nitrogen and phosphorus was polluting lakes, streams and rivers and causing an expanding dead zone in the Gulf of Mexico.

2020 and beyond: Balance, Biology and Brains. I believe that the future of farming will be biological. The present-day agricultural system isn't sustainable. Overuse of agrichemicals, no-till, dead soils, and too many harsh fertilizers are depleting topsoil and impacting soil life, leading to a huge loss in soil carbon.

Biological farming is the future of agriculture. Biological farming means managing farms to promote soil life, using nitrogen more efficiently, using fewer chemicals, and growing healthier, high-yielding crops. In order to be successful biological farmers, we need to get soils healthy and mineralized in order to grow healthy, high quality, disease-resistant crops. This means farmers need to be educated on what to do, and what not to do, to reach that goal.

A Biological Farm Example

A farmer I work with in central Iowa provides an excellent example of how a change in thinking and a change in management can turn a farm around. This farmer came to my meeting a few years ago and said to me, "I have two choices: I can either quit altogether and sell my farm, or I can make some radical changes to turn things around. I'm going broke doing what I'm doing." He had been practicing no-till and planting a corn/bean rotation for the last ten years, and each year his yields were getting lower and his fertilizer and pesticide bills were getting higher. He knew compaction was a problem, and he knew things were not working, but he didn't know where to begin to get out of the mess he was in. After attending my winter meeting, he decided to make a change.

The Iowa farmer went back home, sat down with his wife, and together they laid out a five-year plan for the farm. They studied the Six Rules of Biological Farming and planned how to approach each rule, starting with soil testing. Their goal was to improve soil biology and increase yields by changing tillage systems and reducing chemical and fertilizer use on their farm.

After 10 years of practicing no-till, the first thing the farmer did was go out and subsoil his fields. It took a six-wheeled tractor to pull the subsoiler through that hard, tight ground, and it smelled like a sewer as the subsoiler opened up channels into the soil. The ground was so compacted that no air was getting into it. The soil had become anaerobic and smelled like a swamp.

The next thing the farmer did was take soil tests and add soil correctives to balance the soil nutrients. The soil report showed that he had high magnesium and low calcium, sulfur and trace minerals. He applied lime on some fields, and then put on gypsum (calcium sulfate) and homogenized trace minerals to work towards getting his soils back in balance.

He diversified his rotation by adding cover crops. In the spring he planted oats, a cover crop that sucks up nutrients, improves soil structure, and feeds soil bacteria. He then shallow incorporated the oats back into the soil before planting his cash crop. This released the nutrients held in the oat plant back into the soil, both to feed the soil life and to provide

Loose, crumbly soil with lots of soil life.

nutrients for his next crop. In the fall, he planted cereal rye after soybeans to capture nutrients and to provide a cover on his soil over the winter.

The farmer changed the fertilizers he used, dropping potassium chloride and DAP. He replaced them with balanced fertilizers, which included ammonium sulfate, MAP, potassium sulfate and trace minerals. He also stopped using anhydrous ammonia, and cut back from 175 units of nitrogen per acre to 100 units of nitrogen per acre.

And lastly, he changed his fertilizer application methods. He bought a new corn planter with dry fertilizer boxes, so instead of broadcasting phosphorus and potassium he could apply starter fertilizer right where the crop needed it — down the row.

As a result of these changes, he was able to cut back on his chemical applications. After just two years of biological farming he had cut his chemical bill by 75 percent and at the same time he started to see his yields and crop health improve. Soil magnesium levels dropped and calcium levels rose, and soil organic matter increased by almost 40 percent from 2.7 to 3.6 in just four years. He planted the same variety of corn that he had for years, but he saw an increase in test weights from 52 to 60 pounds per bushel.

He went from an average of 140 bushels of corn/acre to an average of 200 bushels of corn/acre and from an average of 40 bushels of soybeans/acre to an average of 55 bushels of soybeans/acre.

Soon his goal was to grow 300 bushels of corn/acre and 80 bushels of beans/acre. Three years into this process, when he was out in his fields combining beans with the yield monitor going, he reached a spot where the monitor logged 120 bushels per acre. He was so excited he stopped the combine, jumped out, and put flags in the ground to mark that spot so he could go back later to see if he could figure out what it was about that soil in that area that produced such high yields. Later that fall when he was subsoiling that field he reached a point where the subsoiler met almost no resistance. He thought the machine had fallen off his tractor it pulled so easily! Stopping to look around, he realized he was in the flagged spot where he had logged the 120-bushel soybean yields. He now knew why he had such great yields. It was all about soil structure and biology. That spot in that field had beautiful, loose, crumbly soil with an abundance of soil life. In just three years, the Iowa farmer was able to dramatically improve the health of his soils, and turn his farm around from losing money to making a profit.

This is the promise of biological farming — to work with the soils and soil life to reduce chemical use, improve soil and crop health and increase yields. I have seen this Iowa farmer's story repeated all over the Midwest, the country and the world. All it takes is an open mind, a willingness to change, and the ability to step up and take a proactive role in managing your farm.

Organic Farming

A farmer came up to me after a meeting a while back with some questions about organic management. He was transitioning his farm to organic, so that spring he worked up his land and put on 8,000 gallons of liquid dairy manure. Then he planted soybeans. His question for me was, "How do you control weeds when you can't use herbicides?" My answer, "You don't put on 8,000 gallons of liquid manure and then plant soybeans! You just fertilized all the weeds, now how are you going to stop them?"

Organic farming requires more management than conventional farming because in organic farming there are no easy fixes. If you let the weeds get away from you, you can't turn to herbicides to control the problem. You need to have healthy, mineralized soils to grow plants that can resist insects and diseases because if your plants get sick you can't use chemicals to solve the problem. The same goes for livestock. You need to feed a balanced, nutrient-rich ration to keep your animals healthy because you can't rely on antibiotics to fight off diseases.

I farm organically on my own farm, and each and every year is a new learning experience. The weather, for example, presents interesting challenges because I can't let the weeds get away from me. I need to be out cultivating my fields just as soon as the weeds start to germinate. If it's too wet to go out in the fields to cultivate in the week after the weeds start growing, I'll end up battling the weeds all season long.

A couple of years ago a visiting farmer walked out in my cornfield and commented that I seemed to have a few weeds in my corn. There are always a few weeds in my corn, but I don't mind. That's because I generally grow two years of alfalfa followed by one year of corn. If I were running a corn/bean rotation I wouldn't want to leave any weeds in my corn because they would go to seed and then I would have weeds in my beans, which can be hard to control. A few weeds with a corn/alfalfa rotation are not as much of a problem as with corn/beans. As an organic farmer it is pretty difficult to be 100 percent weed free, so I don't aim to be. I adjust my management practices to deal with having a few weeds all the while still growing a healthy crop.

A technique I've experimented with to control weeds in my soybean crop is no-till planting my soybeans with a rye nurse crop. Winter rye planted in the springtime will grow eight inches to a foot tall and then stop, and the soybeans will grow right through it. Two years ago I bulk spread three bushels of rye (which is 180 pounds of rye seed) and two bushels of soybeans per acre. I planted using the fertilizer buggy, but the problem was I ended up with thicker beans right behind the buggy and thinner beans farther away, and the fertilizer spinners were splitting my soybeans. So the next year I decided to try something different. I drilled the soybeans in and then bulk spread the rye. I didn't drill the rye because it would leave too much space between the rye plants. It would have been like planting two alfalfa plants per square foot: what would cover the rest of the ground? It would be solid weeds. By bulk spreading three bushels of rye per acre I get better coverage. I then went out and rotary hoed to get the rye seed covered with soil. After I did this the weather got dry, and no rye came up. You can imagine what the field looked like! I had a field of no-till beans with no chemicals and no weed suppressants. It was a weed patch. I had pretty nice beans in there, but pigweed, lamb's quarters and other weeds surrounded them. If I had let the crop grow and then combined the beans in the fall, all those weeds would have gone to seed and next year it would have been impossible to control the weeds in that field. Instead, I rotovated the field in early summer and worked the weeds and soybeans back into the ground before the weeds could go to seed. Then I planted oats and oilseed radish as a cover crop. I may not have made any money off that field that year, but as an organic farmer, I need to do what's necessary to control the weeds. And it wasn't a total loss. By working in my soybean and weed crop, I put a good dose of nutrients into the soil that will feed the soil life and keep my soils healthy for next year's crop.

I farm 1,000 acres, and if 50 or even 100 of those acres turn into soil-building crops for the year, it's not an economic loss for the farm. I still have enough crops to feed my livestock and make a decent living, and the crop that didn't do well and was worked into the soil this year went to soil building, helping that field produce a bumper crop next year.

The Big Picture

As farmers, we need to step back now and then and examine our role on the farm. Take a look at the big picture. Stand back and look at the land from a distance and see all that goes on and your part in what happens on your land. Notice when the carbons and residues are broken down, and then figure out the best way to incorporate residues and use cover crops to improve soil life. Examine your tillage practices and look for ways to modify how you work your soils to protect soil life while keeping loose, crumbly soil with channels for air and water infiltration. Take some time to review your fertilizer program — what is being used, where and when it is applied, and identify the limiting factors. Evaluate your farm and make decisions from a wider viewpoint. Doing so will help you take steps forward as a biological farmer. Meanwhile, remember that some things take time.

You may not see results this year, but hopefully you will be heading in the right direction.

As farmers, we're trying to do the best we can given our resources, the weather and the rainfall, the soil we have, and the tools on hand. We're trying not to stress the plants in order to allow them to express their genetic potential. You can stress plants both by having an imbalance of minerals and a shortage of minerals — either opens the way for diseases. Our soils have a certain ability to give out minerals and nutrients and hold water, but we can make them better. We can add more carbon to the soil to hold more water, getting it to soak further down when it rains. We can get by with a whole lot less added nitrogen. As our soils get healthier, as the soil systems improve, then we can grow a really good crop, even under less than ideal growing conditions.

Now, there's no way that you're going to be 100 percent successful — no one ever is. You've got to accept the fact that you're going to make the wrong call sometimes. That occasionally you're going to do the wrong thing or use the wrong inputs. You're going to plant under less than ideal conditions or use a product that wasn't really right for your soils or your crop. But if you keep the big picture in mind, you won't get cold feet and go back to saying, "Well, it's only NPK, chemicals and biotechnology that are needed."

One way to evaluate your role as manager is to conduct a thought experiment: What would happen if you simply stuck seeds into the ground on your farm and then stood back? What would grow? If the soil is hard and dead, or blow-sand, there would be a very meager crop, but if it's a rich loamy soil and high in organic matter, it would be a different story. On my farm, if I just stuck seeds into the ground and did nothing else, I could probably get 125 bushels of corn per acre. But if I added compost, and extra potassium, phosphorus, calcium and sulfur; if I address the limiting factors like zinc, copper, manganese (Mn) and boron (B); if I grow a cover crop to capture nutrients and feed soil life, then my farm can produce more. Through management practices I can move my yield up to around 175 bushels of corn per acre. But it doesn't end there. I need to be continually evaluating my farm and my farming practices and asking myself, what more can I do to improve soil health and quality and grow crops sustainably?

In my opinion there are three important steps you can take to improve your role as manager on your farm.

Step one is to look at the **big picture**. Shift the way you look at your farm, and be open to considering alternatives you hadn't tried before. Read, study and attend meetings presenting different ideas in farming. Be open to new ideas and willing to step back and evaluate your farm from a broader perspective.

Step two is to be **observant**. Look for clues on your farm that things are changing, whether those clues come from the plants or the soils. If you try something new one year, don't attribute all of your success or all of your failure to that one thing. For example, say you put fish meal on one field, and at the end of the season the crop on that field is really healthy and has the biggest yield on your farm. Go ahead and apply fish meal again next year, but don't make that your entire program. At that moment, in that situation, it was very successful but that doesn't mean it will work year after year. By the same token, if you apply trace minerals or gypsum or some other new mineral one year and don't see any response, do not give up after one year and never try it again. Particularly in depleted soils, nutrient applications can take three to five years to build up enough in the soil to the point where there is a crop response. There are many components to success on a farm. By monitoring your crops and soils regularly you can look for clues that let you know if your management practices are working to improve things on your farm.

Step three is simply, **basics first**. The basics include:
- regular soil testing
- adding minerals to address limiting factors
- always addressing plant-available calcium
- providing a balanced fertilizer for the soil and the crop you're growing
- tilling to protect soil physical properties and soil life
- avoiding harsh chemicals
- growing cover crops in order to create an ideal home where soil life can thrive and crop roots can grow
- increasing diversity by planting cover crops and rotating crops

Remember, management is doing something about something you can do something about. As a biological farmer I have jobs to do to keep my soil life happy and my crops and livestock healthy. Those jobs include doing everything I can to improve soil and crop health, as well as keeping in mind the big picture and working continuously to improve my role as farm manager. Every year I strive to make improvements on my farm. Sure, there are setbacks — bad weather, insect problems and poor crops do happen. But as long as I keep the big picture in mind, I don't get discouraged. Biological farming is not about finding the one right answer. Rather, I always want to be striving to grow healthy, quality crops in a more sustainable way and managing my land so I'm making improvements every year. That's the fun and the challenge in farming.

Chapter 3

The Six Rules of Biological Farming

Even though every farm is different, there is common ground among all biological farms. No matter what your soil type, what crop you're growing, or what your weather conditions happen to be, there are certain rules that all biological farmers follow. The root of the word biological is "bio" meaning life. Biological farming is all about working with plant, animal and soil life, or as I like to say, bio-logical farming is "logical" common sense farming. Another common ground among biological farmers is that they are dead set against any products that harm biology. Materials that are very acidic or very caustic, have a high salt index, or are known to harm soil biology shouldn't be used. There are also products out there that may have negative side effects on soil life, like genetically modified plants and some chemicals and fertilizers, but at this time we just don't know for sure. Until more is known about the effects of those products, I don't think they should be used. It goes against the goals of biological farming to put anything on the soil that will potentially harm soil life.

As a biological farming consultant I work with all kinds of livestock and every kind of crop you can imagine. I visit the state of Maine to consult on strawberry production; I work with potato farms in Wisconsin, Michigan, and Idaho; I work with dairies, beef producers, and sheep producers in the United States, Australia, New Zealand and South Africa; I work with cranberry farmers, blueberry farmers, wheat farmers, vegetable farmers,

The Six Rules of Biological Farming

1. Test and balance your soils and in addition, feed the crop a balanced, supplemented diet.

2. Use fertilizers which do the least damage to soil life and plant roots. Watch salt and ammonia levels. Use a balance of nutrients, with a balance of soluble to slow release and a controlled pH. Use homogenized micronutrients, add carbon and place them properly to enhance performance.

3. Use pesticides, herbicides, biotechnology and nitrogen in minimum amounts and only when absolutely necessary.

4. Create maximum plant diversity by using green manure crops and tight rotations.

5. Use tillage to control the decay of organic materials and to control soil air and water. Zone tillage, shallow incorporation of residues and deep tillage work great on many farms.

6. Feed the soil life, using carbon from compost, green manures, livestock manures and crop residues. Apply calcium from a good, plant-available source.

corn/bean farmers, and many other kinds of crop farmers. I even work with vineyards in the United States and in Australia.

One of the vineyards I work with in Australia produced an award-winning wine from its biologically farmed grapes. When the wine was tested, the producers said it had the best chemistry of any wine they'd ever seen on their farm. And what do I know about growing grapes? Nothing. Absolutely nothing.

You might wonder how I can work with so many different kinds of crops in so many different areas of the world when soil types and crop nutrient requirements vary so much. Without knowing anything about the crop being grown, how can I help the farmer? What I do is look at a soil test and mineralize the soils based on that soil test. It's true that crops and soils vary, but there are some things that are the same no matter where you farm or what you grow. All crops need a variety of minerals, and all crops need soil biology to grow. I can make recommendations for many different crops because I take soil tests and I ask the right questions. Important questions like, what are you doing now? What's working? Are your soils healthy and mineralized?

The Six Rules of Biological Farming are guidelines I came up with that work for all biological farmers around the globe. They will move you toward your goal of being a successful biological farmer no matter where you live or what you grow.

The Six Rules of Biological Farming

Rule #1: **Test and balance your soils** and in addition feed the crop a balanced supplemented diet. This rule is about understanding your soil's baseline mineral levels, and adding soil correctives to improve that baseline.

Before testing your farm soil, there are a few things about soil testing you should be aware of. First, different soil test labs give different results. Though different labs may conduct the same tests, slight variations in equipment and protocol will affect the results. There are two reasons for always using the same lab: 1) so that soil tests from the same field can be compared from year to year and 2) so that recommendations for soil correctives are consistent.

Second, soil sampling isn't an exact science. Not only is a small sample of soil supposed to represent an entire field, but what is measured from that sample is only what can be extracted in a lab. What the plant can extract from the soil may be different. That's why it's important to also conduct tissue tests or feed tests in addition to soil tests as a way to gather more clues about what nutrients the plant is able to extract from the soil.

Third, soil tests measure soil minerals by extracting them from the sample with a mild acid. Your soil test is limited to measuring what can be extracted by this methodology. Soil tests do not measure biology, structure, or even all of the minerals found in the soil. There are other important factors of soil health that will affect how your soil performs, such as what plants have been grown there, what inputs have been used over the years, how well roots are able to develop, how many earthworms are present, and more. Soil testing just does not tell the whole story!

Another factor to consider is pH. Soil tests are designed to work within a range of pH, and if the pH is high or low, the soil test won't be accurate. If your soil falls into one of these two extremes, plant tissue tests or feed tests are needed to get a better idea of what minerals are available in your soil.

Recognizing that soil testing has its limitations, it's still important to take soil samples every two to five years and use that information to help you improve your soil, and spend your fertilizer dollars where they are needed. A soil test provides you with an estimate of nutrient levels in your soil, and tells you whether the levels are sufficient, in excess or deficient. It's also very important to look at all of the macronutrients and micronutrients, not just NPK. A good soil test includes measurements of phosphorus, potassium, calcium, magnesium, sulfur, zinc, manganese, iron (Fe), copper and boron, as well as organic matter, CEC (cation exchange capacity) and pH.

The first thing to look at on a soil report is what soil correctives are needed. Determining what soil correctives to add is simple. Whatever nutrients are lacking I want to add. Whatever I have enough of, I don't add. On my farm the soils are high in magnesium and low in calcium, so I need to apply calcium but not magnesium. Dolomitic limestone (calcium magnesium carbonate) would be the wrong choice for my farm because I don't need more magnesium. Gypsum (calcium sulfate) or high calcium limestone (calcium carbonate) would be much better choices.

Keep in mind that you need to deal with excesses as well as deficiencies. There are interactions between minerals in the soil, and a mineral in excess can cause as many problems as a mineral that is deficient. In order to provide your crop with a balanced diet, you need to be sure to keep all mineral levels in the soil in balance with each other.

The second thing to look at on your soil report is the crop fertilizer. This is a completely balanced fertilizer for the crop that you are growing that provides nutrients above and beyond what the soil can provide. Even if there is a good balance of all nutrients in the soil, a crop fertilizer is still needed each year to make up for some of the nutrients the crop will remove and to provide trace elements like sulfur and boron that don't build up in the soil.

Rule #2: **Use fertilizers that do the least damage** to soil life and plant roots. Watch the salt and ammonia levels. Use fertilizers with a balance of nutrients, a balance of soluble to slow release and a controlled pH. Use homogenized micronutrients. Add carbon. Place all properly to enhance performance.

Which fertilizer source you use matters. Once you know your soil's deficiencies and excesses based on a soil test, you need to decide the best sources of soil corrective nutrients to use to balance your soil and the best sources of crop fertilizers to use to feed your crop.

There are a lot of types of nutrients available from a variety of sources, and they aren't all equal. Farming is the only industry that I know of where everyone wants to buy the cheapest product without much consideration for quality. No farmer I know would buy a tractor that way! Which would you choose, given the choice between a tractor that you know is just okay and will probably break down a lot — but is cheap, and a tractor that costs twice as much but is very high quality, will last many years and you are confident will not break down? I know I'd choose the more expensive tractor. Why don't you buy your fertilizers the same way? There are a lot of different sources of phosphorus, potassium, calcium and other minerals, and they don't all behave the same way in the soil. What you're after when you buy a fertilizer is performance. That means applying a source of high quality minerals that will promote healthy soils, large root systems and abundant soil life.

There are three main considerations when deciding what fertilizer to use.

First, make sure there are no unwanted elements in your fertilizer. Fertilizer blends that include potassium chloride or calcium chloride contain chloride, a compound that is hard on roots and soil life. You should also check the salt index of your fertilizer source. Many fertilizers contain a high salt index, a sign that that fertilizer source is not ideal for promoting soil life and will be tough on young roots. Try to use fertilizers with a salt index below 100.

☆ NO potassium chloride or calcium chloride   Salt index > 100

Below 100 Better

Fertilizer Material	Salt Index
Potassium chloride	116
Ammonium nitrate	105
Sodium nitrate	100
Urea	75
Ammonium sulfate	69
Calcium nitrate	53
Potassium sulfate	46
Sodium chloride (table salt)	154

Another thing to consider is that some fertilizer sources have a very high or a very low pH. For example, the pH of DAP varies from 9 to 11, way above the 6.5 to 7.5 range ideal for soil life. Adding a fertilizer source with a pH as high as 11 will alter the makeup of soil organisms in the zone where that fertilizer is applied, affect nutrient availability, and interfere with the healthy growth of root hairs. Anhydrous ammonia is another fertilizer that is hard on soil life. It is not the nitrogen that is the problem; it is the concentration and the pH. Anhydrous ammonia can raise the pH up to 12 in the zone where it is applied, and that high pH changes the balance of who will live and who will die in the soil. In addition, adding a highly concentrated dose of nitrogen to the soil from a source like anhydrous ammonia will burn up soil carbon, and its use over time can reduce soil pH by leaching out calcium. There are many better, less harmful sources of nitrogen to use.

Second, use naturally mined materials when possible. Naturally mined materials include sources like high calcium lime and rock phosphate. These sources have not been processed down to their base elements, so they contain traces of nutrients like cobalt and molybdenum that aren't considered essential for plant growth, but which no doubt play an important role in the plant/soil system. It's like eating oats as compared to eating Cheerios. The less processed, "whole food" option is going to provide you with better nutrition. In addition, naturally mined materials tend to be less soluble; meaning the minerals in them will break down in your soil over time and won't leach or run off in a heavy rain. To become plant available, these naturally mined minerals need acidity from the soil, the root zone or soil biology to slowly break them down.

Third, balance the ratio of soluble to slow-release elements. An excess of soluble nutrients can lead to losses through tie up, leaching or runoff. Leaching and runoff are not only an environmental problem, but also mean that nutrients are lost and no longer available for plants to utilize. Too many soluble nutrients can also interfere with soil health and root development. It is important for fertilizers to have a balance of soluble and slow-release nutrients so your crop gets an immediate dose of nutrients up front from the soluble sources, and also has nutrients available to it over the course of a growing season from the slow-release fertilizer.

Rule #3: **Use pesticides, herbicides, biotechnology and nitrogen in minimum amounts** and only when absolutely necessary. Herbicides, insecticides and nitrogen should be applied only when they are needed. These inputs never make things better for the future, and the need for them is a sign that soil health is not optimum. While it may be necessary to use them in the short term, the need for pesticides and applied nitrogen will decrease over time as soil health, and consequently plant health, improves.

One of the biggest concerns with applying soluble nitrogen is runoff. There is a dead zone in the Gulf of Mexico that varies in size each year depending upon how much nitrogen runs off farm fields in the Midwest and heads downriver into the Gulf. With so much nitrogen running off into streams and rivers, it seems that a lot of this expensive material is being applied in excess of what the crop can use and is being wasted. Nitrogen is the only major plant nutrient that you can grow yourself. You can reduce your nitrogen inputs over time by promoting soil life, which increases nitrogen fixation by soil organisms, and by growing nitrogen-fixing cover crops such as clover and hairy vetch.

That said, you can't just go out and reduce the amount of nitrogen you put on your crop without first making some management changes on your farm. You have to earn the right to reduce your nitrogen inputs. One year a farmer who attended one of my meetings decided to cut down on his nitrogen applications, so he bought some of my fertilizer, and went home and bulk spread it on his no-till corn. He's a no-till farmer and without making any changes to any of his farming practices, he slashed his nitrogen inputs by 25 percent. What do you think happened? His yields dropped by 30 bushels an acre. Why wouldn't this happen? He didn't earn the right to reduce his nitrogen. He didn't do anything about those hard, tight soils. He didn't put on any calcium. He didn't put any fertilizer down the row. Before you reduce your nitrogen inputs, you need to grow nitrogen-fixing legume cover crops or succulent cover crops that will feed soil bacteria, apply soluble calcium and sulfur, and get your soils healthy and mineralized so you increase the available nitrogen from your soil system.

Most pesticides have negative side effects. As farmers, we want to minimize the negatives and emphasize the positives. I'm not saying that all pesticides are bad, but I am saying that overuse of pesticides without considering the negatives is a problem. I am against the management strategy of "If it's Tuesday, today I spray." I believe that before using pesticides the farmer needs to consider alternatives and decide whether spraying is absolutely necessary.

My philosophy when applying something to my crop or soils is that it should help me this year and next year. Pesticides may take care of a problem I'm having this year, but they won't help my crop or soils next year, and they may do some harm. We don't know all of the impacts herbicides, insecticides or fungicides have on soil life such as nitrogen-fixing bacteria and soil fungi. Knowing these substances are potentially harmful to soil life, we should always use caution when applying them and never use more than necessary. Cost should not be the major determining factor on whether or not to use these substances — need and consideration of potentially harmful side effects should be.

Rule #4: **Create maximum plant diversity** by using green manure crops and tight rotations. Maintaining a high level of plant diversity on your farm has a lot of advantages. By rotating your crops each year or by including cover crops in the rotation, you feed a variety of soil life so no one group of organisms, including pests and diseases, takes over. Just like a cow prefers eating different plants than a horse does, different types of organisms in the soil prefer different plants. This means that maintaining diversity aboveground will help maintain diversity in the soil as well. Growing an assortment of crops and cover crops also means you cycle a variety of soil nutrients, rather than just the ones your crop uses.

Farmers who grow below-ground crops, such as potatoes, practice tight crop rotations out of necessity, as soil-borne pathogens will build up quickly if the same crop is grown year after year. Farmers don't grow potatoes on potatoes on potatoes; they rotate potatoes with winter wheat and a rye cover crop, or alternate with snap beans and alfalfa. Organic potato farmers in particular often have four or more years between potato crops, because that way the diseases and insects run out of food to live on in between. Maintaining a tight rotation in other crop production systems will have the same benefit.

The more types of plants I grow, the more types of soil life I feed and the better my soils get. Tight rotations are really about maximizing diversity. Every plant species pulls up different types of nutrients from the soil, produces different types of root exudates, and produces different secondary compounds (chemicals plants produce as part of their natural defense system against insects and disease) than every other plant species. It stands to reason that each plant species is going to support or suppress a different group of soil organisms than any other plant species. Therefore, by increasing the diversity of plants you grow, you are also increasing the diversity of organisms in your soil.

If you're growing hay on hay on hay and selling it off your farm, you're exporting nutrients. After selling hay several years in a row, you will start to see a decline in phosphorus, potassium, calcium and other mineral levels in the soil. Different crops remove different nutrients, and growing the same crop over and over depletes the same nutrients each year. While it is possible to fertilize and return nutrients to the soil, increasing plant diversity is added protection against nutrient-depleted soils. In addition, with shorter rotations there is increased diversity of residues. Like the crop itself, different types of residues will have different nutrients in them, both returning a greater diversity of nutrients to the soil and feeding a greater diversity of soil life.

Not all the plant diversity on your farm needs to come from crop rotation. Let's say a farmer wants to grow corn/beans or potatoes/wheat and then to rotate just those two crops. There's not much diversity in that rotation, but the farmer could interseed the corn with clover, or put in rye after soybeans, or grow a fall cover crop of rye and vetch. There are a lot of ways to grow a variety of plants on your farm. On my own farm, I like to rotate two years of alfalfa hay with a year of corn followed by a cover crop blend such as buckwheat/rye/vetch. Sometimes I'll follow that with a year of soybeans or other forage species such as sorghum/sudangrass and sometimes I'll go back to hay. Maintaining a tight crop rotation and growing cover crops keeps disease and insect pressures very low on my farm and allows me to do without pesticides.

Rule #5: **Use tillage to control the decay of organic materials** and to control soil air and water. Zone tillage, shallow incorporation of residues and deep tillage work great on many farms. As a farmer, you have certain jobs to do, key among them is managing air and water and the decay of residues. Till with a purpose, and till as little as possible. Practice what I call "thoughtful tillage."

Aggressive tillage causes problems because incorporating too much air into the soil will burn up carbon and damage soil life and soil structure. It's also important to avoid tilling as much as possible in the middle zone where the crop roots are growing. Once root masses are established, the roots provide pathways for air and water movement into the soil, and the area around the roots, the rhizosphere, is abundant with soil life. Disturbing the root zone tears up root channels, breaks up soil aggregates, and disturbs soil life. Once that happens, it can take a long time for soil life and soil structure to get reestablished.

On the flip side, not tilling at all also presents problems. A high organic matter soil with lots of earthworms and soil life that is never driven on when it's wet or muddy may not require tillage, but that type of situation is rare. I have jobs to do as a farmer, and not doing anything doesn't get them done. Without any tillage to open things up, soils become too tight and a crust forms on top, cutting off air movement into and out of the soil. In addition, without any tillage soils can become compacted, which limits water movement through the soil. This can lead to waterlogging, which causes soils to become anaerobic and a whole host of problems follows.

Soil organisms, particularly the beneficial ones, need to breathe. Like any other farm animal, they can't stick their heads underwater and expect to survive. Therefore, water has to soak in when it rains, and when it is dry, water needs to be able to move back up. That's why I like to sub-soil, opening up deep channels for the movement of water and air in the soil.

In addition to problems with air and water management, no-till also creates problems with the breakdown of residues. Without tillage all residues are left on top of the ground, where most of them will oxidize and their carbon will be given off into the atmosphere rather than being incorporated back into the soil. I like to shallowly incorporate residues, putting a large percentage of them into the top four inches of soil where they are in contact with soil organisms. Some of the residues are left on top for protection from erosion, and some are left stuck in the soil like wicks to aid in air and water movement. The majority of the residues are under the surface of the soil where they can be digested and become part of the soil carbon complex.

When considering what type and how much tillage is appropriate for your farm, always keep in mind your purpose for disturbing the ground and that makes it a lot easier to pick the best course of action.

Rule #6: **Feed the soil life,** using carbon from compost, green manures, livestock manures and crop residues. Apply calcium from a good, plant-available source.

Feeding soil life means providing something more than raw manure and the residues of corn stalks or soybean stubble. I want to feed the soil life a wide variety of things, and this includes living plants. It's important to always keep in mind that the plants determine the soil life, so if you want to increase diversity below the ground, you need to increase diversity above the ground as well.

On my farm, I like to feed my soil life green manure crops, and a variety of them. Some of the green manure crops I use include fall-planted winter rye or a fall blend of rye and vetch, or oats in the spring before planting soybeans. Another way to build up soil life in poor ground or land that is in transition to organic is to take that land out of production for a year for soil building. After the crop is harvested, plant an over-winter cover crop like winter rye, then in the spring plant oats, shallowly incorporate them in summer and plant buckwheat and rye in the fall. The following year you can plant your cash crop or forage crop into soil that will have a more abundant and diverse soil life. You can also build soil diversity by tightening your rotation and including a larger variety of crops, such as for the livestock farmer, a rotation of hay and forage crops.

On a conventional farm the driving force is nitrogen, but on a biological farm it's all about calcium. It's vitally important to understand that calcium is needed for healthy crop production. It is a key mineral that plays a role in both soil and plant health. Calcium's functions are many and diverse. It aids in the uptake of other minerals. It is found in the cell wall of plants as part of pectin, which is useable energy for animals. It promotes soil life. It improves soil structure. It reduces the ability of pathogens to invade cells. And it helps modify soil pH. It is this last important role of calcium in the soil that causes a lot of misunderstandings. The conventional belief about calcium is that if your soil has a high enough pH, it means the plants growing in that soil are getting plenty of calcium. I dispute this belief. Just because your soil pH levels are above 7.0, it doesn't mean there is plant-available calcium in the soil. Not all applied calcium sources are the same, and not all calcium in the soil is accessible to plants. Especially when growing legumes, it's important to apply a plant-available source of calcium each year. Legumes require a lot of calcium, and sources like high-calcium lime are not very plant available. I like to use a source of available calcium on my legume crops each year, such as Bio-Cal or OrganiCal. High calcium lime that has been extremely finely ground, or micronized, is also a good source of plant-available calcium. Gypsum, which contains sulfur as well as calcium, is another good source of calcium, particularly on soils that are high in magnesium. It's important to test your soils and choose the right source of calcium for your situation.

I believe these Six Rules of Biological Farming apply to any biological farmer, no matter what you grow or where you farm. Over the years I have modified them and refined them, but these remain the basic tenets of biological farming. The goal of implementing these rules is to have loose crumbly soils with a good mineral balance and an abundance of soil life. When you reach that goal, you will reap the rewards of healthy plants, healthy animals, and high-yielding crops on your farm.

An example of healthy, loose soil with good air, water and nutrient exchange.

Calcium

Winter rye
spring Oats
Buckwheat/
Rye

Chapter 4

Top 6"

Soil Testing

The first rule of the Six Rules of Biological Farming described in the previous chapter is to test and balance your soils. I believe very strongly that the only way to evaluate the limiting factors in your soil and to make corrections is to conduct regular soil tests. Soil tests have their limitations, but if you recognize those limitations and follow consistent soil sampling protocols soil tests are an excellent tool for improving and maintaining soil fertility.

A soil test gives you a snapshot of what is happening on a sampled field. When you sample the soil, generally 10 to 20 cores (or shovelsful) of soil are taken from the top six inches and mixed together in a bucket. About two cups of that soil is then put in a bag and sent to the testing lab. The top six inches of every acre of land you farm contains about two million pounds of soil, which means that if you take one sample from a five-acre field, the two cups of soil that go to the lab have to represent what is happening on ten million pounds of soil. This is one reason why soil sampling is only an estimation of the mineral levels in your soil. It is not an exact science. The numbers you see on your soil test don't tell you exactly what is happening in your soil, but they do give you a good idea about what minerals are deficient or in excess. The soil test also provides you with clues about how well your soil can hold on to minerals and how much organic matter is present. This information gives you a starting point for making decisions about what soil correctives and crop fertilizers to use, and allows you to monitor the success of your soil fertility program over time.

Composition of Typical Soils*

Components	Sandy Loam (lbs/acre)	Silt Loam (lbs/acre)	Clay Loam (lbs/acre)
Organic matter	20,000	54,000	96,000
Living portion (microbes, earthworms, etc.)	1,000	3,600	4,000
Nitrogen	1,340	3,618	6,432
Silicon dioxide	1,905,000	1,570,000	1,440,000
Aluminum oxide	22,600	190,000	240,000
Iron oxide	17,000	60,000	80,000
Calcium oxide	5,400	6,800	26,000
Magnesium oxide	4,000	10,400	17,000
Phosphate	400	5,200	10,000
Potash	2,600	35,000	40,000
Sulfur trioxide	600	8,500	6,000
Manganese	2,500	2,000	2,000
Zinc	100	220	320
Copper	120	60	60
Molybdenum	40	40	40
Boron	90	130	130
Chlorine	50	200	200

*for a plow layer 6.5 inches in depth, approximately 2,000,000 lbs

Source: J.L. Halbeisen & W.F. Franklin

Soil tests measure minerals by extracting them from the sample with a mild acid. It's important to know that this method does not show the total amount of all of the minerals in the soil. Instead it is meant to be an estimate of the minerals in the soil that a plant can access. There are many more minerals in the soil than are measured by the soil test. Those minerals are tied up in the soil in various ways that make them unavailable to growing plants.

The composition of soils on the previous page shows the total amounts of 15 nutrients found in the top six inches of one acre of topsoil. This isn't all of the minerals as there are closer to 70 different minerals found in the soil. This table shows the minerals known to be needed by a growing plant, which are the ones that are likely to be found on your soil report.

If you send a soil sample in to a lab, your soil report will have much lower numbers than found on this table. That's because the soil report is only telling you the estimated amount of plant-available nutrients, not the total amount of minerals like those shown in the table.

Take a look at the potassium, phosphorus, magnesium and calcium numbers from the table. The top six inches on one acre of a silt loam soil contains a total of 35,000 pounds of potassium, 5,200 pounds of phosphorus, over 10,000 pounds of magnesium and 6,800 pounds of calcium. A soil sample taken from this same silt loam soil and sent to a lab for analysis would have potassium levels of around 200 to 300 pounds per acre, phosphorus levels of around 50 pounds per acre, magnesium levels between 150 and 500 pounds per acre, and calcium levels of around 2,000 pounds per acre. Obviously the soil lab isn't reporting the total amount of these minerals in an acre of soil. Rather, a soil lab estimates what nutrients are active in the soil — in other words, those minerals available to the plant. The lab determines this by measuring the nutrients that go into solution when the sample is subjected to a mild acid. Of the approximately 10,000 pounds of magnesium in one acre of silt loam soil, only around 300 to 500 pounds are in the active part of the soil and show up on a soil test. This means that there is a lot more magnesium in the soil than what is reported on a soil test, but that it is not plant available. It's also interesting to note that a larger percentage of total soil calcium than magnesium will show up on a soil report. The test results for a silt loam soil will show around 25 to 35 percent of the total calcium in the soil, compared to only 3

Soil Nutrients: Total Number vs. Soil Test Amount

	Total amount present	Amount shown on an average soil test
Calcium	6,800 pounds	2,000 pounds
Magnesium	10,400 pounds	150-500 pounds
Phosphorus	5,200 pounds	50 pounds
Potassium	35,000 pounds	200-300 pounds
Organic Matter	54,000 pounds (equal to 2.7%)	2% to 3%

The above table shows the total amount of soil nutrients present compared to the amount of soil nutrients reported on a soil test from one acre of a typical silt loam soil.

to 5 percent of the total magnesium. This is because calcium is more active in the soil so more of the total volume of calcium shows up on a soil test as plant available.

If you look closely at the nutrients listed on the soil test, you will notice that there are no numbers for soil nitrogen. That doesn't mean there's no nitrogen in the soil. The silt loam soils I farm contain about 3,500 pounds of nitrogen per acre, and a clay loam soil would have closer to 4,000 pounds. One reason I do not measure nitrogen on my soil test is because soil nitrogen is a moving target. Over 90 percent of the nitrogen in the soil is organic nitrogen, meaning it is tied to soil life or organic matter and is unavailable to plants. In order for nitrogen to be plant available, soil organisms have to "mineralize" the organic nitrogen into either ammonium or nitrate. But this isn't a one-way street. Just because nitrogen is in a plant-available form doesn't mean it will stay there for long. Available nitrogen can dissipate by moving back into the organic nitrogen pool when soil organisms consume it. It can also leach, volatilize back into the atmosphere, become tied up in the soil, or be taken up by plants. If a soil lab measured plant-available nitrogen levels, by the time the report came back from the lab those numbers wouldn't mean anything because the amount of available nitrogen in the soil would be different. The soil report would tell me what the available nitrogen was when I pulled the soil sample last week, which doesn't help much when planning what fertilizer I need today. That's why I don't measure nitrogen on my soils, or on any of the farms I work with in the Midwestern United States.

One exception to this is in arid regions, such as the Western United States and Western Australia. Where rainfall levels are low, nitrogen doesn't change form as quickly and the nitrogen levels on a soil report are more accurate.

Soil tests, in addition to estimating available minerals, also provide an estimate of organic matter and calculate cation exchange capacity. These numbers are an estimate of how much life is in the soil and the soil's nutrient holding capacity. Though most soil labs measure organic matter, getting a good assessment of organic matter from your soil test is tricky. If you split up a soil sample and send it to five different labs, you will end up with five very different numbers for organic matter. And while the soil report gives you an estimate of how much total organic matter is in your soil, it can't tell you how much of that organic matter is from residues, humus or living organisms.

This is one of several shortcomings of soil tests. Soil tests do not measure soil biology, soil structure, or even all of the minerals found in the soil. There are many important factors of soil health beyond what minerals are

The real secret to biological farming is tapping into the reserve nutrients in the soil through healthy soil biology. One way of doing this is by planting cover crops. Some types of cover crops, like oats and buckwheat, are able to access reserve nutrients that many other plants cannot get.

The added bonus of increasing soil biology is that more nitrogen is produced by nitrogen-fixing bacteria in the soil, as well as by soil life breaking down cover crop residues.

in the soil, such as what plants have been grown in that soil, how well roots are able to develop, how many earthworms are present, and many more. All of these factors affect what nutrients a plant is able to access from the soil. Remember, what is measured from a soil sample is only what can be extracted in a sterile environment inside a lab. What the plant can extract from the soil depends on many chemical, physical and biological factors all interacting in a living, changing environment. This is why I recommend tissue tests or feed tests in addition to soil tests as a way to gather more clues about what nutrients the plant is able to extract from the soil.

The object of soil testing is to get clues about what minerals are deficient in the soil before spending money on soil correctives or crop fertilizers. A soil test can tell you what type of soil you have, the soil's nutrient holding capacity, and what minerals are lacking or in excess. That information can help you make decisions on how to correct limiting factors in your soil and how much fertilizer to use. Soil testing is a tool to help the farmer make decisions. It isn't perfect, but without it farmers would have no idea where to begin.

Different Methods of Soil Sampling

There are three common types of soil sampling. Each one will work for different farmers at different times, and each of the sampling methods has its pros and cons. It really all depends on your goals, your budget, and the equipment you have available.

Whole field: Walk the "bulk of the field" and pull 15 to 20 soil cores. Avoid sampling areas that are different from the majority of the field (i.e. low spots, areas close to a road or stream).

Grid-point: Divide each field into 5-acre sections. Overlay a grid on each five-acre section and pull a soil core from each "square" on the grid. A GPS is used to mark each sample spot.

Management zone: Soil samples are taken within "management zones" delineated by the farmer. Management zones are areas of different soil or crop characteristics within a field. For example, areas of higher and lower yield within a field would be sampled separately to look for differences in fertility in those areas.

Whole Field Sampling

On my farm, whole field sampling fits my goals. My goal is to look at how soil fertility has changed since the fields were last sampled, and to find a fertility program that will work well for a field, or preferably a group of fields. Since the goal is to get a snapshot of soil fertility on a per-field basis, walking the field and collecting cores is a relatively quick and definitely cost-effective way to collect soil samples.

When I sample on my farm I like to start by standing on the edge of the field and looking over it, getting a sense of what the bulk of the field is like. If it's a ten-acre field, there might be four different soil types in it, but because it has been farmed as a single unit most of the field is pretty similar in fertility and production. There might be a couple of knolls or dips, and those are areas I avoid when taking my soil cores. I also avoid edges because it's likely the edges either got more fertilizer or less fertilizer than the rest of the field.

Then I walk the field in a zigzag pattern, avoiding those areas that are different from the majority of the field. I don't ever walk in a straight line because I could end up following along a path that got more fertilizer or manure spread on it, or may have looser soils from tillage or crop roots, or tighter soils if I'm following a wheel path. I go across the field at an angle and take 20 soil cores, making sure that I cover as much of the field as I can while avoiding edges and knolls and such.

One year I had an intern pull my soil samples and they came back extremely high on trace minerals. Based on my previous soil tests, I knew that it was very unlikely that the numbers were accurate. The field was planted to corn when the intern pulled the soil samples, so I asked him a few questions about his methods. He said he had pulled the samples right down the row because that is where the corn was growing so he assumed that I wanted to know the nutrient levels in the area right around the corn roots. A good idea, I guess, except I'd just applied a starter fertilizer down the row. So the soil samples didn't tell me anything about mineral levels on that field, they told me what was in my starter fertilizer!

I always use a soil probe to collect my sample, but a narrow shovel works, too. I sample the top six inches, which makes sense because that's the zone that gets the most tillage, the most fertilizer, and has the most

Soil Testing

How to sample a field

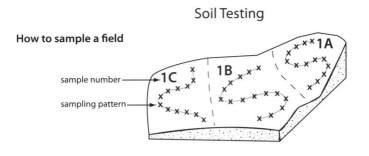

sample number
sampling pattern

How to sample using a soil probe, a soil auger or a shovel

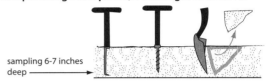

sampling 6-7 inches deep

crop roots in it. A sample from that zone lets me know what is happening in the part of the soil that most impacts the crop. I mix the soil cores all together in a bucket, and then fill my sample bag as evenly as I can with the bigger clods and smaller bits of dirt from the bucket.

I will sometimes go back and pull a second sample from the field if I feel there might be a fertility problem on the knolls or in a dip or other area that isn't like the bulk of the field. But generally I only want one soil sample from each field because I'm not going to buy two mineral blends for one field. I try to keep it as simple as I can.

Grid-point Sampling

Since I do not have variable rate fertilizer application equipment, doing grid sampling generally does not fit my farm. Grid sampling can pinpoint areas of higher and lower fertility within each field, and because each soil sampling spot is recorded on a GPS, you can return to nearly the identical spot the next time you soil sample. This idea is that if you know a field's fertility on a fine scale, you can use variable rate fertilizer application equipment to avoid over-application of crop fertilizers in areas that have enough nutrients, and add more in areas that are deficient. However, grid-point sampling costs more money both in sampling and equipment costs, and it takes more time to do the sampling. Another issue I have with grid sampling is that because of the cost and scale, it's pretty much only used by very large farms that are looking to test pH and optimize their use of phosphorus and potassium. Grid sampling could be used to determine soil correctives on a finer scale as well, but I don't know of any farmers who do that. I like to say that farmers who grid sample to optimize fertilizer application are putting the wrong stuff in the right spot — they know where soil nutrients are lacking, but rather than try to fix the soil in those areas of low fertility, they apply more crop fertilizer there. However, if grid sampling helps farmers reduce inputs of phosphorus and potassium — this is a step in the right direction.

In general, I don't think grid sampling is the best option. I had some experience with this method a few years ago when I worked with a seed farm in my area that grid sampled 160 acres for phosphorus and potassium levels and applied variable amounts of those nutrients based on the test results. They measured yields in the fall and then re-sampled the exact sample grid to see if there was a correlation between phosphorus and potassium levels and yields. What they found was that the highest yielding spots often had the lowest levels of soil phosphorus and potassium, and the areas with the highest levels of phosphorus and potassium were often the lowest yielding. The seed farmer thought these results made sense because at the highest yielding spots the crop took off the most nutrients from the soil. That meant the nutrients had moved from the soil to the plant in those areas, so the soil was low in nutrients. Okay, but what about the areas with the highest soil levels of phosphorus and potassium and the lowest yields? My question to the seed farmer was this, "does that mean that if we raise the phosphorus and potassium levels of the lowest testing spots to a higher level, we can uniformly reduce yield across the field?"

Management Zone Sampling

This is a very good way of soil sampling if you have problem areas within your field that you want to check out. For example, if there is an area of consistently lower yield within one field, by doing management zone sampling you can figure out whether that area has a lower pH, lower calcium levels, lower CEC, or some other issue like poor biology and poor

Veris Machine Field Map

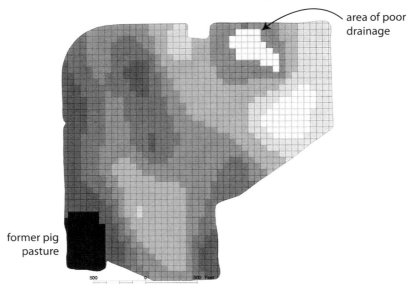

area of poor drainage

former pig pasture

soil structure. You can then apply soil correctives to that one area of the field to improve fertility and help fix whatever else is broken.

A form of management zone sampling I tried on my farm was sampling using a Veris machine. A Veris machine runs a disc through the ground to measure electrical conductivity, and produces a map of soil texture, along with soil water-holding and nutrient-holding capacity in 50 foot strips across a field. I used the Veris machine to create a soil texture map from a 120-acre field on my farm. After the map was produced, soil samples were pulled from 26 locations in the field shown to be different from one another. The sample data was put together with the Veris machine data to create a map of soil texture, soil nutrients, CEC and pH across the field. The map showed an interesting range of nutrient levels and pH.

The Veris machine pulled out two areas in the field where the soil pH or nutrient levels were very different from the rest of the field. The first spot that was highlighted is an area that's been a problem on that field for years. It's a water hole. One spring I subsoiled that spot, which drained the water. The next year it was a water hole again. When we tested the soil with the Veris machine that spot stood out as different, and the soil sample showed that it had a 5.5 pH. Until I get a calcium soil corrective into the soil in that spot I will continue to have problems with soil structure and drainage, so I will continue to subsoil and add lime to that spot each year until I get the problem fixed.

The other interesting thing we found with the Veris machine was an area of high fertility in the southwest corner of the farm. We hadn't done anything different with that part of the field, so I went to the former owner and asked him what that corner of the farm was used for. He said that 50 years ago it was a hog pasture. The nutrients put into the soil from years of hog manure still showed up on a soil test five decades later! Even though that field has been farmed as one field continuously for the 50 years since it was a hog pasture, that corner still has higher phosphorus, copper, iron and zinc levels than the rest of the field. It goes to show that as long as you continue to feed the crop, once the soil is fixed, it's fixed.

I also think it's beneficial to take subsoil samples in problem areas on a field. Rather than sampling the top six inches of soil, sample from six inches down to 12 inches. You may find that the subsoil has a low pH and is low on calcium, which would explain why the soil isn't draining well in that area. Or maybe the subsoil is depleted of potassium, which would be a clue as to why your alfalfa crop isn't doing great. A subsoil sample can provide you with more clues about fertility on your farm. You then need to decide what you'll do to fix the problem.

Unlike tilled fields where the nutrients get mixed into the top six inches of soil, on pastures and no-till fields the nutrients stay on top. In those cases, taking a soil sample from the top six inches isn't going to be a good representation of what's happening. Instead, sample the top two to three inches of soil, and then the top six inches and compare them. This will give you an idea of how the nutrients are moving in the soil. It may be that 90 percent of the soil fertility is in the top two inches — and at that point you would want to consider incorporating nutrients deeper into the soil. When things get dry in the summer, the first thing to dry out is the top layer of the soil, so if your nutrients are all in that dry soil, they aren't going to be plant accessible. If the soil is low in fertility below that top layer of concentrated nutrients, the plants will run out of nutrients as the soil dries out. Or if there is a heavy rainfall, the nutrients in that top layer could wash away. It's very important to get those nutrients deeper into the soil where roots can access them.

Along with soil sampling for nutrient analysis, you should also do some digging around in your soils to get a sense of soil quality. How good is root development? Are the roots flattened out and growing horizontally, or do they grow straight down deep into the soil? Is your soil blocky or crumbly? How difficult is it to dig a 12-inch deep hole? If it has been dry the soil will be difficult to dig into, but if you sample when the soil is moist you can get a good idea of how tight or loose and crumbly it is.

Take a notebook and paper out to the field with you and take notes on what you see. Keep a file on each of your fields that you can look at each time you soil sample to see if you notice changes. The goal of soil sampling is to improve soil quality, and keeping good notes will help you monitor changes in your soils.

What Lab to Use

Different soil test labs give different results. Though different labs may conduct the same tests, slight variations in equipment and protocol will affect the results. It doesn't really matter which lab you use, as long as it is the same lab each year. I always send my soil samples to Midwest Laboratories out of Nebraska for two reasons: 1) so soil tests from the same field can be compared from year to year; and 2) so recommendations for soil correctives and crop fertilizers are consistent, based on soil sampling results that I know are reliable.

I always use Midwest Labs when I'm working in the central part of the United States, but when I'm traveling I often use different labs. The most important thing is that once you start working with a lab, you need to stick with it. You can't change horses in the middle of the stream. Even if a different lab is using the same procedures as your usual lab, you will get slightly different results on your soil test that don't reflect anything that is going on in the soil. This could result in applying nutrients that you do not need, or not applying nutrients that you do need.

Remember, soil sampling is not an exact science, but it is a very important tool for understanding and improving soil fertility. There are quite a few different methods for taking soil samples. The important thing is to be consistent with what you do on your farm. Always follow the same soil sampling protocols and always use the same soil testing lab so your results will be consistent and you can compare them from year to year and from field to field. By monitoring your soils and taking regular soil samples, you can continue to improve the quality of your soils and the health of your crops.

Chapter 5

Interpreting the Soil Test

How to Evaluate Your Soil Test Results

After you take your soil samples and send them in to the lab, one day you'll get an envelope in the mail. It's your soil test results. You look it over and compare your mineral levels to the "desired levels" written on the report. Some things look high, some low, and some average. Now what?

Below is a soil test from a farm I purchased a few years ago. I took soil samples before I planted anything, added anything, or began doing any soil building. The following table (see next page) is what the soil report showed.

Phosphorus and potassium levels look high, magnesium is a little high but not too bad, calcium is a little low but also not too bad. Organic matter and CEC levels are what they are, but what do those numbers mean? And how will those numbers help me determine what soil correctives and fertilizers I need? Using my farm as an example, the following is an explanation of the numbers on the soil report.

Prepared for: Otter Creek Organic Farm

Field	OM	CEC	pH	K+ ppm	Mg+ ppm	Ca+ ppm	Na+ ppm	P1- ppm	P2- ppm	S- ppm	Zn+ ppm	Mn+ ppm	Fe+ ppm	Cu+ ppm	B- ppm
											Micronutrients				
1	2.2	12.4	7.2	235	397	1683	20	132	169	17	3.9	15	55	2.1	0.6
2	2.3	10.8	7.0	334	277	1505	14	172	192	13	6.1	19	43	2.2	0.6
3	2.4	10.8	7.0	319	271	1527	13	152	193	12	5.1	17	35	1.8	0.5
Desired level:			6.8	--	--	--	--	50	100	50	5	20	20	2	2

Base Saturation:	% H+	% K+	% Mg+	% Ca+	% Na+
Field 1	0	4.9	26.7	67.7	0.7
Field 2	0	7.9	21.4	70.1	0.6
Field 3	0	7.6	20.9	71.0	0.5
Desired level:	--	3-5	12-16	70-75	--

Soil type

Soil type is evaluated by looking at CEC and organic matter (OM).

Soil type	CEC	Where the Otter Creek soils are
Light soil	5-8	
Medium soil	8-15	Fields 1, 2 and 3
Heavy soil	15-22	
Extra heavy soil	22+	

Cation exchange capacity, or CEC, is a measure of your soil's ability to hold and release nutrients. Clay particles and humus in your soil have a negative charge and can hold on to positively charged minerals. These positively charged minerals are called *cations*, and the more clay and humus in your soil, the more cations your soil can hold. Thus, the higher your CEC, the greater your soil's ability to hold nutrients. Think of the CEC like a dinner plate. The larger the plate, the more food (nutrients) it can hold, but this also means that if your plate is empty (i.e. your nutrient levels are low), it takes a lot of food to fill it back up.

CEC: Cation Exchange Capacity

Cation exchange capacity is the ability of a soil to adsorb and exchange cations. "Adsorption" just means that something is loosely held, similar to the way something with static electricity, like a sock or drier sheet, is loosely held by your clothes.

Clay particles and stable organic matter (humus) both have many negative sites on their surface that can adsorb nutrients. Because opposites attract, these negative sites hold on to positively charged molecules, called cations.

CEC varies between soil types. The more clay and humus a soil has, the higher its CEC, and thus the greater its ability to hold on to cations. In practical terms, this means that some soils have

---

Lots of negative sites to hold cations
Humus + Clay for greater CEC

Soil Texture and Cation Exchange Capacity (CEC)

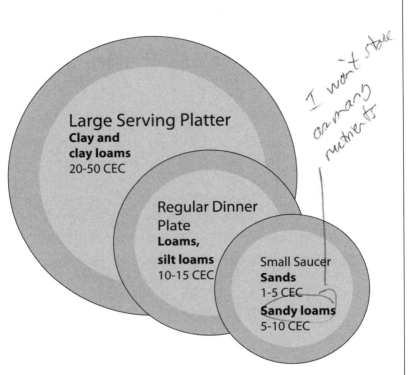

Large Serving Platter
Clay and clay loams
20-50 CEC

I won't store as many nutrients

Regular Dinner Plate
Loams, silt loams
10-15 CEC

Small Saucer
Sands
1-5 CEC
Sandy loams
5-10 CEC

a greater capacity to store nutrients, while other soils have more limitations.

The analogy of a dinner plate works well for understanding CEC. The higher the soil's CEC, the larger the "plate" the soil has for holding soil nutrients. A soil with a lower CEC has a smaller plate, meaning that soil cannot hold as many nutrients. But just because the soil has a high CEC, or large dinner plate, doesn't necessarily mean that all the right nutrients are in that soil available for plant uptake. Just because your plate is the largest, doesn't mean you have the most food on it, or a good mix of different types of food. The person next to you with a smaller plate could have more types of food on his small plate than you have on your large plate. Like CEC, the plate just represents the holding capacity not the contents.

When nutrients are removed from the soil and not replenished, the dinner plate gets low on food or imbalanced in the types of food it holds, such as having only potatoes and no meat or vegetables. This is one reason regular soil testing is essential. A soil test can tell you what nutrients are depleted so the dinner plate, your soil, can be refilled with what it is lacking.

Remember, it is not only the amounts of soil minerals that are important, but also the balance of minerals. The amount of a particular nutrient a plant can access can be affected by the proportion of that nutrient in relation to other nutrients in the soil. It's very important to have a good balance of nutrients in the soil to achieve both good yields and high quality crops.

Cation Exchange On Clays

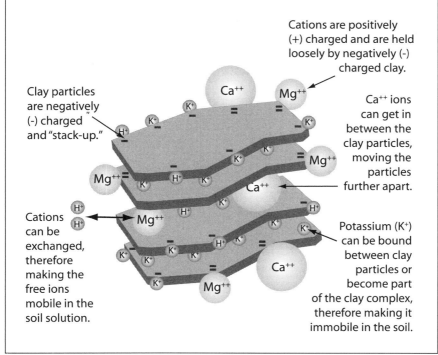

Cations are positively (+) charged and are held loosely by negatively (-) charged clay.

Clay particles are negatively (-) charged and "stack-up."

Ca^{++} ions can get in between the clay particles, moving the particles further apart.

Cations can be exchanged, therefore making the free ions mobile in the soil solution.

Potassium (K^+) can be bound between clay particles or become part of the clay complex, therefore making it immobile in the soil.

Base saturation measures how full your plate is, and what's on the plate. Base saturation is the percentage of exchangeable bases, or cations (potassium, magnesium, calcium, sodium and hydrogen) in your soil. The total of all these cations has to equal 100 percent. You know your plate is low on food (ie: your soil is low on nutrients) if hydrogen makes up a significant percentage of your base saturation. Hydrogen takes the place of nutrients, so when hydrogen levels are high, another nutrient has to be low. Hydrogen also contributes to a low pH (see discussion of pH that follows), so another way of knowing your soil has low fertility is if the pH is low.

Organic matter is a measure of how much carbon is in your soil, both the living plant and animal portion such as that in roots and earthworms, and the dead and decomposing portion. Organic matter is extremely beneficial to growing good crops. Generally speaking, as long as drainage is good the higher your organic matter levels, the healthier your soils. Levels below three percent can be cause for concern, but this also depends on your soil type (and the lab where the samples were tested). Lighter soils, those with a low CEC, tend to have lower organic matter levels.

Reading the Otter Creek Organic Farm soil test for organic matter: All three fields in this example are between 2.2 and 2.4. Because this is a medium soil with CECs between 10 and 12, organic matter levels are on the low side, but not too bad.

Minerals & pH

pH

Soil tests are designed to work within a range of pH. If the pH is unusually high or low the soil test results will not be accurate. If your soil falls into one of these extremes, plant tissue tests or feed tests are essential in order to get a clear picture of what nutrients plants are able to access from the soil.

A good range for pH is generally between 6.8 and 7.2. This is the range where both soil fungi and soil bacteria live (though fungi prefer a lower pH than bacteria, and this affects the balance of organisms in your soil). Also, in this pH range there is a balance between maximizing the soil minerals held by clay and humus in your soil, and the ability of plants to access those minerals.

If the soil pH is under 6.5, lime is recommended to bring the pH closer to 7.0. How much and what kind depends upon your soil type, what other minerals are present in the soil, and how quickly you want to move the soil pH. On a low pH soil, soil test results for minerals are often artificially high, and these numbers will come down after liming. Also, at low pH the availability of iron and aluminum to the plant go up at the expense of other generally more desirable cations such as calcium and magnesium. Unless you're growing a specialty crop like blueberries, a low pH is not beneficial to the plant's health, energy or nutrition.

At a pH over 7.5, soil test numbers may be artificially low and may not accurately reflect the level of nutrients plants are able to access in the soil. Phosphorus is often tied up in high pH soils, and phosphorus deficient plants lack energy and are unhealthy. In addition, a high pH will affect soil life, and can mean the soil has a lower fungal population. One way of mitigating this problem is to add organic matter to the soil by growing green manure crops and amending the soil with compost and manure.

Reading the Otter Creek Organic Farm soil test pH level: All fields in this example are between 7.0 and 7.2, which is a good range for getting accurate soil test results and maximizing the minerals held on the clay and humus in the soil. This is also a good pH range for earthworms and other soil life to thrive.

Phosphorus

P1 level	P2 level	Otter Creek Organic Farm P1 level	Otter Creek Organic Farm P2 level
High: 50+	High: 100+	132–152	169–193
Medium: 20-50	Medium: 40-100		
Low: below 20	Low: below 40		

Phosphorus provides energy to plants, so it's good to have sufficient levels in soils. However, phosphorus inputs are carefully regulated because runoff of this nutrient is a source of waterway pollution. Careful management of phosphorus by working inputs into the soil minimizes runoff. Once phosphorus is in the soil, it tends to tie up rather than to leach. It's also very difficult to move phosphorus levels in the soil. When they're high they tend to stay high, and when they're low they tend to stay low. It takes a lot of inputs to bring phosphorus levels up. I like to see phosphorus at 50 ppm of P1 (the amount of plant-available phosphorus in your soil) and 100 ppm of P2 (the reserve amount of phosphorus in your soil). Other labs or consultants are likely to recommended a P1 of 25 and a P2 of 50. These lower P1 and P2 levels do represent the amount of phosphorus adequate for plant growth, but I like to see higher numbers. Phosphorus is used by plants to store energy, and having extra available phosphorus in the soil means plants can take up more phosphorus, improving their overall health and vitality.

A word of caution: just because there are adequate phosphorus levels on your soil test does not mean there is adequate plant-available phosphorus. It's difficult to get good phosphorus uptake without soil organic matter and healthy soil life. For example, a light soil with a CEC of five or less and low levels of organic matter that shows high levels of phosphorus on the soil test can still leave your crop short of phosphorus. The best way to know if your crop has adequate phosphorus is to take a tissue test or a feed test. A tissue test reading that shows high phosphorus is a very good indicator of a healthy, biologically active soil.

Reading the Otter Creek Organic Farm soil test phosphorus levels: The phosphorus levels, both P1 and P2, are very high. In this sample P1 is 132-172 and P2 is 169-193. As long as phosphorus levels remain high, only small amounts of phosphorus should be applied to these fields. I like to see phosphorus at 50 ppm of P1 (the amount of plant-available phosphorus in your soil) and 100 ppm of P2 (the reserve amount of phosphorus in your soil).

Potassium

Potassium + Sandy Soil

For potassium, the soil test will show two numbers: one result in ppm (parts per million) and the other in percent. Soil CEC levels affect the relationship between percent and ppm. Lighter soils (those with a lower CEC) will have a higher percent of potassium at a particular ppm than heavier soils. For example, a light soil with five percent potassium might have only 75 ppm, while a heavy soil with only three percent potassium might have 200 ppm. Because the percent of potassium in the soil varies so widely based on soil type, the range of values considered high, medium or low for potassium will depend on your soil type.

I like to see potassium values over 125 ppm and above 2.5 percent in the soil. If the soil is close to pure sand, the percent potassium required to reach 125 ppm is very high, and it is difficult for the soil to hold onto that much of the nutrient. On sandy soils I recommend spoon-feeding potassium to the crop throughout the growing season in the fertilizer to ensure good crop production. In addition, some specialty crops require high levels of potassium, which can be applied as a sidedress.

Unlike phosphorus, potassium levels in the soil change easily and an excess can cause problems. High potassium levels can indicate a history of high nutrient inputs, while low numbers can indicate low inputs, low CEC, or a history of exporting forages. While it's important to have adequate potassium to grow your crop, too much potassium uptake by forages can cause animal health problems.

Reading the Otter Creek Organic Farm soil test potassium levels: This farm has medium soils, so potassium levels of five to eight percent and 235-334 ppm are high. Until potassium levels come down, very little potassium should be applied to these fields.

Calcium

Calcium is extremely important for growing good legumes. Just like corn and other grasses need plenty of nitrogen to thrive, legumes need plenty of calcium. Calcium also improves soil structure, and is very important for plant growth and health. The relationship between calcium, magnesium and potassium has an influence on soil structure and plant uptake of calcium. I like to see 70 to 75 percent calcium in a healthy soil.

Calcium w/ Sulfur

Reading the Otter Creek Organic Farm soil test calcium levels: Calcium levels on these fields range between 67 and 71 percent, which means they are at the lower end of the ideal range for calcium. I like to see 70 to 75 percent calcium. A good soluble calcium source with sulfur in it should be added to these fields each year.

Magnesium

Magnesium is part of the plant's chlorophyll molecule, which makes it extremely important for growth and energy production. It is also important for protein production and enzyme function in plants. While it is very important to have sufficient magnesium in the soil, on the flip side, too much magnesium can cause soils to become tight and can limit plant uptake of calcium and potassium. I like to see 12 to 20 percent magnesium on a healthy soil.

Reading the Otter Creek Organic Farm soil test magnesium levels: Magnesium levels in this example range from 21 to 27 percent. This is high, which means a soil corrective containing calcium and sulfur is needed to help lower magnesium levels and should be applied each year. With magnesium levels this high it will be difficult to overdo sulfur, and the addition of sulfur will increase the amount of plant-available magnesium. I like magnesium between 12 to 20 percent.

Sulfur

Sulfur is integral to protein production in plants, so it is very important to have sufficient amounts in the soil. Sulfur is an anion (negatively charged particle), so it doesn't adhere to soil particles and is prone to leaching. Low sulfur levels (below 15 ppm) indicate that sulfur should be added. High sulfur levels (above 20 ppm) can indicate subsoil compaction, low rainfall, or a recent sulfur application. Because sulfur leaches, it always needs to be part of your fertilizer program. I recommend a minimum of 25 pounds of sulfur in sulfate form ("sulfate sulfur" means in a form like calcium sulfate, ammonium sulfate, magnesium sulfate, potassium sulfate, etc.), per acre per year.

Reading the Otter Creek Organic Farm soil test sulfur levels: Sulfur levels are low, ranging from 12 to 17 ppm on the fields in this example. I like to see sulfur levels between 15 and 20 ppm. A minimum of 25 lbs./acre of sulfur should be added as part of the crop fertilizer each year. Because soil magnesium levels are high, adding extra sulfur will provide not only more sulfur to the crop, but also more plant-available magnesium.

Micronutrients

Micronutrients are essential to plant growth, health and reproduction, as are the macronutrients. They are called "micronutrients" not because they are any less important than macronutrients, but because they are found in small quantities in the soil and in plants.

Boron — Add yearly — It leaches

Boron, like sulfur, is an anion and prone to leaching. It aids calcium utilization in plants — so insufficient amounts can lead to plant growth and health problems. If your boron levels are below two ppm, I recommend adding it to your fertilizer blend. As with sulfur, boron needs to be a part of your annual fertilizer program. I recommend one pound of actual boron per acre per year. Sodium borate is a good source of this micronutrient. The source I use contains 15 percent boron, so seven lbs./acre should be applied to reach one lb./acre of actual boron.

Reading the Otter Creek Organic Farm soil test boron levels: Boron ranges from 0.5 to 0.6 ppm on the fields in this example. Levels below two ppm are low. Some boron should be included in the crop fertilizer each year. With the lighter soils and relatively high rainfall on this farm, one pound per acre of actual boron is the minimum that should be applied each year.

Zinc — Relationship of phosphorus + zinc

Zinc is very important for photosynthesis, so it is important to maintain adequate amounts of this mineral in your soil. If zinc falls below five ppm I recommend adding more. Zinc levels also need to be higher if phosphorus levels are very high. A 10:1 ratio of phosphorus to zinc is ideal. If the ratio gets higher than that (meaning there is more than 10 times more phosphorus than zinc in the soil) it's difficult for the plant to get enough zinc, and more needs to be added to the soil.

Reading the Otter Creek Organic Farm soil test zinc levels: In this example, zinc ranges from 3.9 to 6.1 ppm, which is a medium to adequate amount. Levels below five ppm are low. However, given the high levels of phosphorus on these fields the ratio of phosphorus to zinc is close to 30:1. This high ratio of phosphorus to zinc means that plants will have difficulty accessing the zinc that is in the soil, and zinc should be added each year as part of the crop fertilizer.

Manganese & fungus

Manganese plays a crucial role in photosynthesis, and is also important for nitrogen translocation and enzyme function in plants. It is involved in the production of lignin in plants, and sufficient amounts can help protect plants against attack by fungal disease. Manganese availability is closely tied to pH. If the soil pH is above 7.2, manganese availability decreases. However, if a tissue test shows good levels of manganese in the plant on a high pH soil — that is a good indicator that the soil has healthy biological activity. The type of test used to measure soil manganese varies from soil testing lab to soil testing lab, so the amount of manganese considered adequate will vary depending on what lab you use. On soil samples sent to the soil testing lab I use, Midwest Labs, I like to see manganese levels at approximately 20 ppm.

Reading the Otter Creek Organic Farm soil test manganese levels: Manganese ranges from 15 to 19 ppm in this example, so levels are on the low side. I like levels at approximately 20 ppm, so manganese should be added to the crop fertilizer each year.

Copper

Copper is involved with the immune system of plants. It is important for proper enzyme function, and can help control molds and fungus. It also increases stalk strength and helps produce a higher quality crop with greater storability. I like to see copper levels over five ppm.

Reading the Otter Creek Organic Farm soil test copper levels: Levels in this example range from 1.8 to 2.2 ppm. This is a pretty good amount of copper; however some should be added to the crop fertilizer each year to be certain the crop is accessing enough of this micronutrient. Ideally, I like to see copper levels over five ppm.

Iron

Nitrogen-fixing bacteria in the soil require iron. It is also important for chlorophyll production and energy release in cells. Adequate amounts of iron in the soil are necessary, but if levels get too high it can tie up phosphorus. This is a problem if iron levels are higher than P2 phosphorus levels on a soil test. If iron is deficient (below 20 ppm), there is the potential for chlorosis and decreased photosynthesis in plants.

Reading the Otter Creek Organic Farm soil test iron levels: Iron levels in this example range from 35 to 55 ppm. These are on the high side, but they are much lower than P2 levels, so phosphorus tie up is not a concern. However, no iron should be applied to the fields. I like to see iron levels at approximately 20 ppm.

The Farm's History

Before I can make recommendations, I need to understand a few things about the farm. Looking at soil test results gives me clues, but I also need to know more about the farm's history, things like:

What types of crops were grown on the field?

Have any pesticides been used?

What types of fertilizer or lime have been applied?

Does the farmer have a manure source?

The farm used in this example, one of three farms that make up Otter Creek Organic Farm, is right across the road from the main dairy. Several years ago, the people who were farming this land before I bought it saw me combining corn on our field across the road, and one of the farmers came running over and asked what my corn yields looked like. I said, "Jump in!" so he climbed into the combine and rode with me for a while. The yield monitor was bouncing between 190 and 220 bushels per acre as we travelled across the field. He was shocked. His field directly across the road from mine was a Pioneer corn test plot, and it averaged 145 bushels per acre. His land had been farmed conventionally for years as a corn/bean rotation with lots of applied manure, and he had expected his yields to be comparable to mine. Given how much lower his yields were than those on my field across the road, I had some idea that this land wasn't in the best condition physically and biologically before I bought it.

I also knew that the people farming this land before me had a laying hen operation and a beef feedlot. This farm is on level ground only a couple of miles away from a 100,000 hen shed, so in addition to manure from the beef feedlot, it got lots and lots of chicken manure. I knew that the farmer used 9-23-30 as a starter fertilizer, knifed in anhydrous ammonia, and used pesticides to control weeds and insects. The farm was planted to a corn/bean rotation for many years with no cover crops or any other variation in the rotation. Before I ever dug into that soil, I had an idea that it would be hard ground with little soil life and very high levels of phosphorus and potassium from all of the applied manure.

Name:

Date:

Prepared by:

Soil Fertility Recommendations

Field Numbers & Acres		SOIL CORRECTIVES		CROP FERTILIZATION
	High Testing P&K Soils: MAIN CONCERNS:	1st Year: 2nd Year: 3rd Year:	*Start with Ca & P*	*Balance concentration recovery*
	Medium Testing P& K Soils: MAIN CONCERNS:	1st Year: 2nd Year: 3rd Year:	*Use natural rocks*	
	Low Testing P&K Soils: MAIN CONCERNS:	1st Year: 2nd Year: 3rd Year:		

Recommendations Based on the Otter Creek Organic Farm Soil Test

When making recommendations, I like to use the previous table to group fields as high, medium or low testing. Even though soil correctives involve more than phosphorus and potassium, I group fields by phosphorus and potassium numbers because that is a good starting place for making soil corrections. Also, if phosphorus and potassium levels are low, I know other nutrients will need correctives as well. I start by balancing phosphorus, generally by applying rock phosphate and livestock manures if they are available. I then balance soil calcium by finding a good calcium source that fits the farm based on the soil report.

On this farm I only need minimal soil correction because the soil nutrient levels are pretty good and I don't need to do any major fixes. Soil phosphorus levels are very high so no soil corrective phosphorus is needed. The soil test shows adequate calcium in the soil, and calcium to magnesium ratios that are almost ideal. Most of the farms in this area have high magnesium levels and low calcium levels in the soil — about 30 percent magnesium and 50 percent calcium is typical — but this farm has about 20 percent magnesium and close to 70 percent calcium. Even so, I like to apply a good calcium soil corrective each year. Gypsum fits the farm well, so I applied 1,000 pounds per acre to all of the fields.

After I've addressed soil correctives, I start to look at what nutrients my crop needs this year; i.e. the crop fertilizer. My approach to crop fertilizers is different than what is done in the conventional world. Instead of replacing phosphorus and potassium as the crop removes them, I apply a balanced fertilizer that fits the farm. Because my farm is organic and I have different rules to follow, I don't use many soluble nutrients. Most of the nutrients in my crop fertilizer come from manure, compost and naturally mined sources. My crop fertilizer provides a balance of nutrients above and beyond what the soil can provide.

On this example farm, potassium levels are 4.9 to 7.6 percent and phosphorus P1 is around 150 ppm on all three fields. I would group all three fields under "high testing" on my Soil Fertility Recommendations table and formulate a crop fertilizer with only a small amount of phosphorus and potassium in it. For nutrient management purposes, having high levels of phosphorus and potassium on these fields is a problem. I have dairy cattle and lots of manure on my home farm right across the road and it's not wise to spread too much of it on these fields. I'm going to have to haul the manure somewhere else.

The adequate levels of calcium in the soil on this farm are likely due to all of the chicken manure added over the years. Laying hens are fed four percent calcium in their diet, and a lot of it comes right back out in the manure, which is eight to twelve percent calcium. Because the calcium has been through a chicken before it gets on the fields, it is a good source of available calcium. Even with adequate levels of calcium on the soil report, I will still apply a good soluble calcium source as part of my crop fertilizer each year.

Micronutrient levels on this farm are low, and I doubt the former owners applied any. Sulfur and boron levels are especially low. Since those nutrients are anions, they don't attach to soil clay particles, which means they will leach, making it very difficult to build up levels of these two nutrients in the soil. Zinc levels are five to six percent, which looks good. However, because of the high phosphorus levels in the soil, getting zinc into the plants will be difficult. Remember, a ten to one ratio of zinc to phosphorus is ideal. With P1 levels at around 150 ppm on this farm, I need 15 ppm of zinc so I'll need to add extra zinc to my fertilizer. Copper levels are around 2 ppm, which is good, but I will still add some each year to be certain the plants are able to access enough. A good homogenized blend of trace minerals mixed with a carbon source like humates and some gypsum will work really well on these fields.

The crop fertilizer I formulated for this farm includes calcium, sulfur, carbon (in the form of humates), and the trace minerals zinc, manganese, copper and boron. As a result of the high soil fertility, I applied only 200 to 250 pounds per acre of starter fertilizer along with a low rate of compost and manures from the dairy. I planted corn that first year and placed the fertilizer blend in the row near the root zone.

I also really need to work on soil biology, which I see as the biggest limiting factor to good yields on this example farm. A cover crop in the fall like winter rye and maybe an early spring oat crop would work well. Mechanical soil aeration with a ripper should also be beneficial. Getting more air and water into the soil will help stimulate soil biology.

Based solely on nutrient levels, the farm in this example should be very high yielding, but in our first year farming this land we really struggled. During the first spring on the new farm I sent one of my farm workers over to rotovate the fields across the road after I had finished rotovating on the home farm. I was planting on the home farm for most of the morning, and across the road from where I was I could see his tractor. He only had this tiny narrow strip done. I thought he was taking a nap. So I went over there and told him we have 1,000 acres to farm, you can't spend all day on one field! He said the tractor didn't have enough power to pull the rotovator. There were mudballs coming off the back of the rotovator and it was barely moving through the soil. If a 175 horsepower tractor couldn't pull a 10-foot wide rotovator, that was some hard, tight ground — now I understand why the neighbors only got 145 bushel per acre corn yields. You see, just looking at the soil test doesn't tell the whole story. *A soil test measures chemistry, but it is not only minerals like calcium and magnesium that influence soil structure, it is also biology.* The ground was as hard as a stone when we tried to plant that first spring, and we didn't get great yields that fall. There was way too much foxtail and other weeds that we were not able to control and certainly hurt our yields. I should have just subsoiled, added my soil correctives, and planted green manure crops for the first year to build soil biology before planting crops.

For biological farmers, coming up with soil nutrient requirements requires knowledge not only of nutrient levels, but also knowledge of what's been grown on the soil, what types of inputs have been used, how much soil life is present and many other factors. It isn't a numbers game. That's one reason why it is so helpful to work with a good, knowledgeable consultant, especially when you're starting out in biological farming. Most conventional farmers and consultants just look at what nutrients the crop removes and add those back to the soil each year — in many cases adding only phosphorus and potassium. But because of the complex interactions between soil life, growing plants, and minerals in the soil, biological farmers don't add nutrients just based on the numbers. There is a lot more to it than that. Soil life plays a critical role in making nutrients available to plants, and a soil test is never going to be able to take that into account. The soil test can give you clues about what nutrients are in the soil and the soil's capacity to dish out those nutrients to your crop. A soil test is a useful tool to help the biological farmer select soil correctives and a good crop fertilizer, but it doesn't tell the whole story. It is important to remember that biology also plays a central role in providing nutrients to growing plants and keeping plants healthy.

Chapter 6

Tissue Tests & Feed Tests

Most farmers take soil tests to see which minerals are lacking so they can make decisions on what nutrients they need to add in order to grow a good crop. But just because nutrients are added to the soil is no guarantee they'll get into the plant. The soil is a complex environment, and not all of the nutrients found in the soil are available to plants. Taking regular tissue tests on row crops and feed tests on forage crops is a way to check on how well your fertility program is working and whether your plants are accessing the nutrients they need. They are another tool that provides you with clues on the health of your soils and plants. If the soil has major limiting factors it will show up in the plant, because even in healthy mineralized soils not all nutrients are plant available. Nutrients can be tied up or inaccessible, in which case nutrient deficiencies won't show up on a soil test. Testing the plant is the only way to verify that the plant has all the nutrients it needs to be healthy.

As a dairy farmer, I have another important reason to test my forage crops. It is essential for me to know what my cows are eating so that I can be sure they're getting a balanced ration and will stay healthy. I want all of my forage crops to be nutrient-dense, and to be consistent from field to field and crop to crop so I can keep my herd healthy.

Tissue Testing

How and when you take tissue tests will vary depending on what crop you're growing. Different plants have different nutrient requirements throughout their growth cycle and need to be tested at different stages. Ideally, I would test all of my forage crops before each cutting, and my corn and bean crops twice each year. Some farmers test even more frequently than that. For example, most potato farmers take biweekly petiole tests. A petiole test is a plant tissue test that measures nutrients in the part of the plant that attaches the leaf to the stem. If the petiole test shows a nutrient deficiency, farmers can then go to their fields right away and add the needed nutrient. Potato farmers test a lot more frequently than other farmers because of the type of soils they farm and the high value of their crop. For most farms, mine included, taking tissue tests isn't about responding immediately to a change in nutrients from a single tissue sample. Instead, it is about looking for trends and patterns over time, providing clues on how well plants are accessing nutrients in the soil.

Ideally, I recommend taking tissue samples from a corn crop twice a year. I like to take the first tissue test on my farm each year in early June when the corn is healthy, the leaves are green and opened, and the plants are growing so fast you can almost see them grow. This is also a great time of year to be out monitoring the crop. I dig in the soil, pull out a whole plant and look at how the roots are doing, and get a general sense of the health of my soils and my crop. In June the corn plant is young, so I cut the whole plant off at the base and send it in to the lab for a mineral analysis. It is very important not to get any dirt in the sample or the tissue test results won't be accurate. The results of this early season mineral analysis let me know if my crop is short of any nutrients early in its growth cycle, and it also shows ratios between minerals in the plant so I can monitor this as well.

I like to take my second tissue test in mid-summer when the plants are tall, dark green, and healthy, and putting most of their energy towards flower and seed production. This is a point where the plant is using a lot of nutrients, and a tissue test can give me clues as to why my crops aren't producing up to their maximum potential. For example, if I see that zinc levels are low on one tissue test, I know my plants are having difficulty accessing zinc at that time. I won't respond with a change in nutrients based on just that one tissue test, but if I see low zinc levels on the next tissue test as well, when weather and soil conditions are different, I know I probably do have a deficiency. I will then add more zinc to the homogenized trace minerals in my crop fertilizer.

It is very important to take tissue tests under ideal conditions. If it is too dry or too wet, or if the crop doesn't look good and the plants are clearly unhealthy, you don't need to test to see if something went wrong — you already know something is wrong. A tissue test from an unhealthy plant will most likely show a lack of minerals, but that shortage may not be due to a lack of available nutrients in the soil. Maybe the roots were eaten by insects. Or maybe the pith is clogged and rotten from disease. Maybe you put too much of a salt fertilizer down the row, and it burned off some of the roots and damaged the plants enough that insects attacked the crop. It could also be a lack of soil biology that is leading to hard soils that stunt root growth and limit the plant's ability to access nutrients. The problem in these cases is not poor fertility, so you would not want to spend money on fertilizer to correct a problem that was caused by something else. A tissue test would show mineral deficiencies, but it wouldn't show the cause of the problem. When the crop is unhealthy, it is often difficult to tell what is at the root of the problem.

I want to know if my crop is getting enough minerals under ideal conditions. If I take tissue tests when the plants are growing and look healthy, my tissue test gives me a good picture of what nutrients the crop is able to access from the soil. If the tissue test shows deficiencies, I know I have a nutrient uptake problem. This could lead to other problems like insect attack or disease later on, so I want to find any mineral deficiencies before any other problems appear.

It is important to remember that if a nutrient is shown to be lacking on a tissue test, adding more of that nutrient to the soil might solve your problem, and it might not. Just because a mineral is added to the soil doesn't mean it can get into the plant. There is a lot going on in the soil with nutrient interactions, soil structure, and soil biology that can make nutrients more or less plant available. Tissue testing is a tool to let you know how your soils are doing and if the plant is able to get all of the minerals it needs. But adding needed nutrients based on a tissue test might not be enough. You may have difficulty getting those nutrients into the plant due to a lack of soil biology or poor soil structure. A poor tissue test lets you know that you have limiting factors. Discovering what they really are, and then addressing those limiting factors, requires detective work on your part to find the true root of the problem.

The ultimate goal of tissue testing is to help you do a better job as manager of your farm. It will give you clues as to whether or not you're heading in the right direction. It will also give you confidence that when you spend money on fertilizers you are heading in the right direction and things will get better in the future. I believe that tissue testing — whether it is on corn, soybeans, cotton, bananas or rice, is an important part of the fertility program on any farm.

Feed Testing

On my farm, I test every hay crop at every cutting. If I take five cuttings of hay off of a field, I will get five feed tests from that field, which gives me a good picture of what's happening there over time. There is a correlation between the soil test and the feed test, but it's not 100 percent. If my soil tests say I have enough of a certain mineral and my feed tests show it is lacking, I have a feeling the plant isn't lying. The plant isn't getting enough of that mineral, and I know I need to make changes to make that mineral more plant-available.

By testing every hay cutting, I also get the benefit of testing my crop under different weather conditions. Sometimes the soils are dry; sometimes they're very wet. I know that when it is dry certain minerals, like potassium, will be at lower levels in the plant. Testing consistently each year allows me to look for larger trends in mineral levels in my crop and make changes to my fertility program based on those larger trends rather than what I see at each individual cutting. Just like with soil tests, there are limitations to what feed tests can show you. Feed testing gives you a good idea of what nutrients are in your crop, but again, it isn't an exact science. If the numbers on my feed test seem out of line, it could be a sampling error, either on my end or by the lab. If I suspect a sampling problem, I will either pull another sample from that crop, or wait to see what the next feed test shows before making changes based on that test. Remember, the purpose of feed testing is to show trends and patterns over time, not to give black and white, hard and fast answers.

One reason I measure my feed is because I want all of my soil tests and all of my feed tests to look the same. I don't want some fields higher in nutrients than others because then I'll be feeding my cows inconsistently, sometimes with higher nutrient hay and sometimes with lower nutrient hay. That change in feed quality makes it more difficult to balance my ration and manage herd health. I want to consistently feed my cows nutrient-rich feed, and I want all of that feed to be similar in quality and nutrients.

Feed testing is an excellent tool for monitoring what I feed my animals, and it is also a good tool for monitoring my soil fertility, as feed tests also show how many nutrients my crop is removing from a field. Although replenishing nutrients is not a numbers game, by monitoring feed tests and yields across my farm, I will know if some fields need more manure and minerals added than others.

An Average of Four Feed Tests From One Hay Field at the Otter Creek Organic Farm (Taken During One Growing Season)

Test	Average on my fields	Target amount	7 ton crop removes
Crude Protein	22.6	18-21	--
Nitrogen	3.6		504#
Nitrogen:Sulfur	10.6:1	10:1 – 12:1	
Protein solubility	65.1		
Fat	3.97		
Ash	9.6%	7 – 13%	
Lignin	7.3%	5 – 11%	
Calcium	1.45%	1.5%+	203#
Phosphorus	0.33%	0.35%+	46.2#
Magnesium	0.31%	0.35%+	43#
Potassium	1.71%	1.5 – 2.0%	240#
Sulfur	0.34%	1:10 S:N	48#
Manganese	55 ppm	>35 ppm	0.77#
RFV	162		
RFQ	190		

The example feed test from Otter Creek Organic Farm is an average from testing four cuttings of hay off of one field in a single growing season. The four hay cuttings averaged 22.6 percent protein, which is 3.6 percent nitrogen. Given my 7-ton forage yields over the growing season this means that 500 pounds of nitrogen left that hay field and went into the barn. How much nitrogen did I buy? None. Some of the nitrogen in my crop came from manure and compost, and I grew the rest. A healthy, biologically active soil will fix atmospheric nitrogen, and a healthy alfalfa plant can produce up to 200 lbs./acre of nitrogen through nitrogen fixation in the root nodules.

This hay is 22.6 percent protein, which is rich feed for the cows. The reason this hay is so high in protein is in part because it was cut early. Younger alfalfa plants are higher in protein than older plants. I farm with my son, and he is always looking for ways to improve milk production. Recently I told him that if he wants to get more milk from his cows, he has to improve the quality of his forages. To him, part of that equation is cutting early and getting more protein in the feed. So about the middle of May he had the discbine hooked up and the tractor idling, and he was just waiting for me to say, "Go ahead." Instead I said, "It's your farm!" and that was all I had to say and he was off. And that is how we ended up with such high protein hay.

As an aside, here's a funny story about my son; last year my son bought a new truck but kept driving the old one. He let the new one sit in the shed because he didn't want to get it dirty. My father was the exact same way! I got the son my father wanted. Years ago my father bought a brand new Oliver 88 tractor. It was a wet spring, so my father didn't drive his new tractor all spring because he didn't want to get it dirty. He just drove his old one until the fields dried out. My son is just like him — it's funny how genetics works.

Many farmers focus on replenishing the potassium and phosphorus removed by their hay crop, but a good hay crop removes a lot of calcium from the soil as well, and that needs to be replaced, too. Calcium on my feed averaged 1.56 percent, which is 249 pounds of calcium removed for every acre of hay ground. I like to apply OrganiCal or Bio-Cal to replace that calcium. A ton of Bio-Cal contains close to 175 pounds of soluble calcium. By applying a ton of Bio-Cal per acre I can replenish most of the soluble calcium my hay crop removed from the soil.

Bio-Cal for soluble Calcium

> ## Would applying a ton of high calcium lime supply enough calcium to replace the 249 pounds of calcium removed by the hay crop?
>
> A ton of high calcium lime contains close to 700 pounds of calcium, but most of the calcium in high calcium lime isn't available to plants. In one ton of high calcium lime, only about two to five pounds of the calcium is soluble, meaning it's immediately available for the plant to take up. Applying a ton of high calcium lime per acre will help correct a low pH and will move calcium levels in the soil in the long term. But unless you have a low pH soil, it won't supply much calcium to your hay crop the year you apply it.

After seeing some of my feed test results, a farmer commented to me that he could not get his calcium levels in his hay anywhere near where mine are. I asked him what calcium source he was applying, and he said he put on 200 pounds of gypsum per acre, which sounds good, but 200 pounds of gypsum contains only about 40 pounds of calcium. My hay crop removed 250 pounds of calcium in one season! Calcium is the most important nutrient for growing healthy legumes. Just like grasses such as corn need nitrogen to thrive; legumes like alfalfa need calcium to thrive. Without enough soluble calcium on your alfalfa crop, you aren't going to get high yields, healthy plants, or good quality feed.

This hay averaged .33 percent phosphorus, which is 46 pounds of phosphorus removed on each acre. On one of my fields, I removed 69 pounds of phosphorus (reported as P, not P_2O_5) in my hay from one acre in one year. The soil test from that field showed that I had 100 pounds per acre of available phosphorus. Does that mean the next year's soil test will read only 31 pounds per acre of phosphorus (the 100 pounds per acre shown on my soil test minus the 69 pounds per acre removed by my hay crop)? No, because I have a reserve of around 5,000 pounds of phosphorus in my soil (see the previous soil testing chapter for a discussion on the amount of total phosphorus vs. available phosphorus in the soil). This reserve is the amount of phosphorus in the soil that doesn't show up on a soil test because it isn't plant available, at least not yet. But through healthy biological activity I can tap into that reserve phosphorus. We know that biology plays an important role in phosphorus availability. A microbiologist once told me that the only way to double the phosphorus levels in your feed is through soil biology. You can apply all the phosphorus fertilizer you want but you'll never get that phosphorus into your feed if you don't have life in your soil. Applied phosphorus ties up in the soil very quickly, and the best way to make it plant available again is through soil biology.

I can't make up for all of the phosphorus my crop removed with what is in the soil, I also need to be replacing some of the soluble phosphorus my crop removes each year. I have a dairy farm so there is plenty of manure to apply to my alfalfa crop. If the hay ground is too far away for me to haul manure, I apply rock phosphate. I like to blend rock phosphate with an acidic carbon source, like compost or humates, to help make that phosphorus more available to my crop.

Manganese is another nutrient I monitor in my feed. This hay averaged 55 ppm, which in my opinion is an excellent number. Similar to phosphorus, manganese uptake is linked to biological activity in the soil. If you have a feed test coming off of neutral pH soils, and the test shows high manganese along with low iron and aluminum levels, that means there is biological activity in that soil. In my opinion manganese and other trace minerals keep cows healthy and get cows bred. In my dad's time, farmers would feed oats to their dry cows for that same reason. Today, we know that oats take up a lot of manganese, and I'm convinced manganese is linked to both good fertility in cows and good fertility in the soil.

From the feed test results, potassium is on the low side in this hay. Dry soils, older alfalfa stands, or even just limited available soil potassium may be the problem. I will need to add more potassium to my crop fertilizer next year.

Phosphorus, magnesium and sulfur are in a 1:1:1 ratio on this feed test. Why do so many farmers put on phosphorus and potassium, and ignore the sulfur and magnesium? If the hay crop is removing equal amounts of each of these minerals, I want to apply a balanced fertilizer to make sure the crop has enough of all the minerals it needs, not just some of the minerals it needs.

I take feed tests to monitor the amount of minerals in my forage crops for three reasons. First, I want as many minerals in my feed as possible. I want to provide my cattle with mineralized, high quality, highly digestible forages so that they are getting most of their minerals from their feed rather than from a mineral supplement. Second, I want every field to have similar levels of minerals so that each one produces similar quality feed. Providing my cows with a consistent diet helps me keep my herd healthy. And third, forages are high-yielding crops that remove a lot of minerals from the soil. I want to be sure I'm not mining minerals out of my soil, but replenishing those minerals my crop removed so that my soil quality and crop quality stay high.

Maintaining soil mineral levels is one of the reasons I take tissue tests from my row crops, but it isn't the only reason. Because I keep residues on the land and work them back into the soil, my row crops don't remove as many nutrients from the soil as my forage crops. Taking regular tissue tests from all of my row crops helps me figure out what minerals the plants are able to access in the soil, and which ones they aren't, so I can reduce limiting factors on plant health and yield.

A soil test gives you a picture of what minerals are in the soil, but it doesn't tell you what minerals your plants can access. The only way to know what minerals the plant is getting is to take a tissue test or a feed test. Just because minerals are in the soil is no guarantee they're getting into the plant. There are a lot of complex interactions that occur that can make minerals more or less available to plants. Testing the plant is the only way to know for sure that it has all the minerals it needs to grow and be healthy. Taking a tissue test or a feed test is one more tool that allows me to reduce my limiting factors so I can grow a healthy, high-yielding crop. Like any type of testing, tissue tests and feed tests have their limitations, but if you are aware of those limitations, testing plants is a very important tool to help you improve fertility and yields on your farm.

Chapter 7

Soil Correctives
& Crop Fertilizers

Separating soil corrections and crop fertilizer is a wise move on your part. They aren't the same thing, and they each play an important role on your farm. The difference is simple; soil correctives are minerals added to the soil in order to "fix" the soil, while crop fertilizers are nutrients applied each year to feed the crop.

Most soils are not in balance, meaning they lack some nutrients and have too many of others. Soil correctives are used to fix those imbalances. How do you know if your soil has imbalances? A soil test. When I get a soil report back, the first thing I do is look at the levels of the minerals on my soil report and decide what soil corrections are needed. I follow two basic principles. If a nutrient is in short supply I add more. If there is enough I don't add any more. For example, if your soil test is high in magnesium don't add any more of it, but if the soil test is short on calcium you need to apply some.

The purpose of applying soil correctives is to add minerals in order to reduce limiting factors and to balance the soil. A soil balanced physically, biologically and chemically has healthier, more abundant in soil life and produces healthier crops. Applying soil correctives moves your soils toward being balanced chemically. It is only one leg of the chemical/physical/biological stool, but each leg is dependent upon the others. All three legs have to be there in order to have healthy soils.

Three-legged Stool

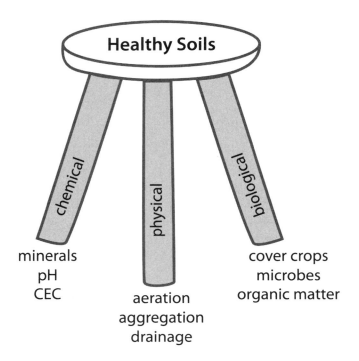

Healthy Soils

chemical — minerals / pH / CEC

physical — aeration / aggregation / drainage

biological — cover crops / microbes / organic matter

Once I've added my soil correctives, then I want to put on a crop fertilizer. I always apply a balanced crop fertilizer that includes some soluble and some slow-release minerals as well as trace minerals. To ensure maximum plant uptake of my crop fertilizer, I want to apply it near or in the row where the plant roots can easily access the applied nutrients.

Many farmers will ask, "If I have applied soil correctives and my soil test shows I have balanced soils with good nutrient levels, then why do I need a crop fertilizer?" The soil has a certain ability to dish out nutrients, but adding a crop fertilizer each year helps maximize yield potential. Also, some minerals, like sulfur, boron and nitrogen, leach from the soil and need to be added each year. I want to be sure my crop isn't starved of nutrients. By adding both a soil corrective and a crop fertilizer I balance my soils while also feeding my crop.

Field Numbers & Acres		SOIL CORRECTIVES		CROP FERTILIZATION
	High Testing P&K Soils: MAIN CONCERNS:	1st Year: 2nd Year: 3rd Year:	*Start with Ca & P*	*Balance concentration recovery*
	Medium Testing P& K Soils: MAIN CONCERNS:	1st Year: 2nd Year: 3rd Year:	*Use natural rocks*	
	Low Testing P&K Soils: MAIN CONCERNS:	1st Year: 2nd Year: 3rd Year:		

Soil Correctives

When I get a soil report, I look it over and the first thing I do is to group fields based on high testing levels, medium testing levels or low testing levels of phosphorus and potassium. The reason I group fields on phosphorus and potassium and not calcium or magnesium or another indicator is for manure management. Grouping them by phosphorus and potassium gives me a quick way of assessing which fields need more manure, which fields are doing fine, and which fields need to be backed off of manure. My goal is to get all of the fields looking exactly the same on a soil test. The same general approach can be used for crop farmers who don't have access to manure. Rather than balancing the soil for phosphorus and potassium based on manure inputs, crop farmers can balance the soil using soil corrective minerals, preferably from naturally mined sources.

If you are a dairy farmer, having the soil tests from all of your fields look the same is very important for maintaining herd health. For example, one dairy farmer I work with ran an eight-year crop rotation. He had one field in corn for eight years, the other field in hay. Manure was his main input, and he put all of his manure on the cornfield and put nothing on the hay field. After eight years, he rotated his fields, seeding his cornfields down to hay, and planting the hay fields to corn. What do you think this did to the nutrient levels in his crops? All of a sudden he went from feeding low phosphorus and potassium hay to feeding really, really high phosphorus and potassium hay when his hay moved to the heavily fertilized former cornfield. If the nutritionist comes out and tests last year's hay, nutrient levels in the hay will be very low. By the time that feed test comes back, the farmer will probably have used up that old hay and have moved on to feeding this year's hay, which is going to have very high phosphorus and potassium levels. The problem now is that the ration was balanced based on what the cows used to eat (the low testing feed), not what they're eating right now (this year's high testing feed). That's a recipe for cow trouble. That's why as dairy farmers we want every soil sample looking the same; then and only then can we have a consistent feed supply for our cattle.

If you're a crop farmer, having the soil test from every field look the same isn't going to be an issue of feed consistency; it will be an issue of mineral management. Most farmers I know don't want to mess around with a lot of different fertilizer blends. It is much more efficient to get one semi-truckload of fertilizer that can be used on all fields. The easiest way to ensure that all of your fields look the same on a soil test is to use soil correctives to balance your soils. If all of your fields have the same ability to dish out nutrients, one fertilizer blend applied across the farm can provide what the crop needs. This helps you to reduce limiting factors on yield, and also makes life easier because one fertilizer blend works for the whole farm.

How I Changed My Soils Over 8 Years

Before I buy a farm I always want to know what shape the soils are in. Before I bought our main dairy farm, I spent some time with the farmer asking him what his farming practices had been over the past 40 years. It is one of the easiest ways to find out if the soils are depleted and will require a large investment in soil correctives. In this case, the farmer told me that the last time he put manure on the soils was just before he sold his dairy herd, about 20 years ago. He'd been growing corn and beans since, and applying 50 pounds of urea and 50 pounds of potash per acre each spring. He also said that he hadn't applied lime in around 35 years. That

Soil Changes at Otter Creek

Prepared for: Otter Creek Organic Farm

											Micronutrients				
Year	OM	CEC	pH	K+ ppm	Mg+ ppm	Ca+ Ppm	Na+ ppm	P1- ppm	P2- ppm	S- ppm	Zn+ ppm	Mn+ ppm	Fe+ ppm	Cu+ ppm	B- ppm
0	1.9	8.9	6.9	78	355	1180	22	37	9		1.7	25	32	0.5	0.7
2	1.9	8.8	7.2	99	295	1125	45	84	8		2.4	16	30	1.1	0.8
8	2.0	10.3	6.9	81	271	1506	55	96	27		2.8	7.3	28	1.2	1.0
Desired level:			6.8	--	--	--	--	50	100	50	5	20	20	2	2

Base Saturation:

	% H+	% K+	% Mg+	% Ca+	% Na+
Year 0	-	2.2	32	65	
Year 2	-	2.9	28	69	
Year 8	-	2.0	22	73	
Desired level:	-	3-5	12-16	70-75	--

This Otter Creek Organic Farm soil report shows how the soil on this farm changed over eight years. Year 0 is the soil test from before I bought the farm. Year 2 is the soil report after I had added soil correctives, rotated crops and planted cover crops for two years, and year 8 is the soil report after I had run the farm for 8 years.

information by itself gave me a pretty good idea of what shape the soils were in, and gave me an idea of what soil correctives were needed. I knew the soils would need lime, and I knew that they were going to be low in phosphorus and micronutrients.

If I know what condition the soils are in, why should I spend time and money on soil tests? There are several good reasons. First, I want some kind of evidence about the condition of the soils when I first start working with them. Anecdotal evidence from the farmer gives me a rough idea of what the soils need, but a soil test tells me where the soil is at right now and how far I need to go to correct it. On this, my example farm, I knew the soil hadn't been limed for 35 years, but did that mean soil calcium levels were just low, or were they very, very low? And what about pH? Just because the soil had not been limed, it doesn't necessarily mean the pH was low. I need a soil test to give me more specific answers to my questions about the condition of the soils. Second, as I start working with the soils and adding soil correctives, soils tests are a way I can monitor whether my soil correctives are working and let me know if I need to make adjustments. And finally, soil tests let me know the soil's shortcomings so I can add extra nutrients to the crop fertilizer to make sure nutrients are not a limiting factor for my current year's crop.

Looking at the Year 0 soil test, I would rate this as a medium soil based on the CEC. The pH is neutral, which I like to see, but the soil is short on nutrients, with phosphorus and potassium levels being especially low. On a medium soil like this I don't like to see potassium levels below 125 ppm, and this soil had only 78 ppm of potassium, while phosphorus P1 was only 22 (about half of what I like to see). Magnesium levels were high and calcium was low, which is typical for soils in this area, and also consistent with what the farmer told me about his calcium applications. Micronutrient levels were also quite low, especially sulfur, zinc, copper and boron.

How to decide what soil correctives to use? It's really quite simple. Whatever nutrient you are short — you add, whatever you have enough of — you don't add. How much to add will depend on your budget and how fast you want to fix the soil.

Calcium and Phosphorus: I always start by correcting with phosphorus and calcium. These two minerals are the slowest to change in the soil, so

I want to add them right away, and in large enough amounts that I will begin to see a difference in soil levels in a couple of years. When I bought the main dairy farm, P1 (available phosphorus) was at 22 ppm and calcium was at 65 percent. Both of those minerals were short so I wanted to add more of them, as I like P1 to be around 50 ppm and calcium around 70 percent.

I wanted to fix the soils as quickly as possible, so I applied a heavy dose of soil correctives that first fall. Not all farmers take this approach — some prefer to add a smaller dose of correctives over a number of years. As long as you don't expect big changes right away, the slow-but-steady approach is easier on the budget, and still gets the job done.

To correct the low calcium and phosphorus levels in the soil, I added 1,000 pounds of high calcium lime, 1,500 pounds of Bio-Cal (Midwestern Bio-Ag's proprietary calcium blend), and 700 pounds of Idaho rock phosphate to each acre. I like to use manure as a soil corrective for phosphorus, but the first two years I owned this farm I didn't have any dairy cows, so instead I used rock phosphate. (Rock phosphate is an excellent soil corrective because it is a source of slow-release phosphorus, and it also supplies calcium and trace nutrients, something I'll discuss further in the next chapter.) Once I got dairy cows on this farm, I applied manure to these fields in the fall.

Now look at how the soil test changed two years later, and then eight years after I bought the farm. Two years into the program, calcium went up four percent to a total base saturation of 69 percent, and after eight years calcium went up a total of eight percent to a total base saturation of 73 percent. Eight years after the initial large calcium corrective was applied, followed only by a soluble calcium crop fertilizer each year after, the calcium levels look really good. In two years phosphorus went to 45 ppm, and after eight years it climbed to 55 ppm. Those numbers are both P1, which is the available phosphorus in the soil. I like to see P1 at around 50, so my soil phosphorus levels are also now where I like to see them in a good, balanced soil.

pH: The starting pH on this farm was 6.9. That's a number I like to see because when the pH is seven you know you have filled all of the spaces on your clay/humus complex with positively charged minerals. If the pH is lower than seven, there are places on the clay/humus complex that have hydrogen attached to them rather than minerals, limiting your soil's mineral holding capacity. I also like to have a pH of seven because earthworms and other soil life do well at this level.

A pH that is either high or low will affect which soil correctives I choose. At a low pH, less than 6.5, the soil is slightly acidic which means I can apply a less soluble mineral like rock phosphate and the acidity in the soil will break it down into plant-available nutrients. If the pH is above 7.5, rock phosphate will break down very, very slowly, making it a poor option as a soil corrective under those conditions. However, mixing the rock phosphate with an acidic carbon source like mined humates before applying it can make it more available, even in higher pH soils (something discussed in more detail in a later chapter). At low pH I will add high calcium limestone as a soil corrective because it is a good calcium source that will also move the soil pH up closer to seven.

In this case, the calcium and phosphorus sources I used did not affect the pH. After eight years of working with these soils, it remained at 6.9.

Magnesium: Magnesium is very, very high so why would I want to add more? I like to target around 15 to 20 percent magnesium in my soils, and on year 0 this soil already had 32 percent magnesium. This is a good example of a nutrient that is in excess, and I need to formulate my soil corrective to reduce the amount of magnesium and improve the ratio of magnesium to calcium in the soil. To lower magnesium levels I added sulfur. Sulfur takes magnesium out of the soil by combining with it to form Epsom salts, (magnesium sulfate), which is highly soluble. This means it is more plant available, but it will also leach from the soil. On this farm, sulfur came from the calcium sulfate (gypsum), potassium sulfate and trace minerals (zinc sulfate, manganese sulfate, and copper sulfate) that I applied. I prefer using the sulfate form of minerals because it gives me the calcium, potassium or trace mineral I want, plus plenty of sulfur. If my farm was not organic, I would also apply a little ammonium sulfate.

In the Midwestern U.S., where my farm is located, soils are typically high in magnesium and most of the farms I work with need to reduce the magnesium levels in the soil. But I certainly don't want to give the impression that magnesium is bad and needs to be reduced in all soils. Magnesium is an essential nutrient needed for plant growth and photosynthesis and I need plant-available magnesium in my soil; what I don't need is an excess of reserve magnesium. A lot of farmers in the Western U.S., for example, are short of magnesium and need more. In their situation, magnesium needs to be added to the soil as part of a soil corrective.

After two years of applying soil correctives on this farm, magnesium dropped to 295 ppm and after eight years it fell to 271 ppm. The magnesium dropped ten percent over eight years, from 33 to 22 percent of my soil cations. Adding calcium, potassium and other cations to the soil, and not adding any magnesium, changed these percentages in part because the percentage of other cations went up in relation to the percentage of magnesium. I like to see magnesium levels around 18 to 20 percent because with the sulfur I add as part of my crop fertilizer program each year, I am always losing some magnesium. It works well for me to have a little extra magnesium in the soil, because a deficiency would cause problems. I also don't want soil magnesium levels to be so low that I need to cut back on my sulfur application because that would cause other problems.

Potassium: If you look at the potassium levels you will notice that over eight years they went from 2.2 to 1.9 percent. The potassium level was low to begin with, and after eight years the level dropped even farther. Why? This happened because I did not apply any soil corrective potassium. I can row apply potassium as part of my crop fertilizer to get enough for the plants without changing the soil potassium levels. I don't often apply potassium as a soil corrective because I don't want to interfere with calcium and magnesium uptake. Calcium, magnesium and potassium are all cations, and they are on a teeter-totter when it comes to plant uptake. If the plant has lots of potassium available to it in the soil, it will take up more potassium and less of the other cations, and I don't want my plants to be short of magnesium or calcium.

I may not apply potassium as a soil corrective, but I do address it in my crop fertilizer program each year. Before I brought cattle onto this farm, I applied 200 pounds per acre of potassium sulfate each year. Once the dairy cows arrived, manure became my potassium source.

Trace Minerals: The trace mineral levels on this farm didn't uniformly move up. In part, this is because it's very difficult to build sulfur or boron levels in the soil. They're anions, which means they will leach because they are negatively charged and aren't held by clay or humus in the soil. If you look at the soil test, it looks like sulfur levels increased drastically between

> Sulfur, boron and nitrate nitrogen are not added as soil correctives because they're anions and they leach. Phosphorus is an anion, too, so how is it possible to do a soil corrective for phosphorus?
>
> When phosphorus from manure, rock phosphate, or another phosphorus fertilizer source is added to the soil, it is in the form of phosphate, which has a strong negative charge. Because of this strong negative charge, phosphate will bind to calcium, magnesium, aluminum or iron in the soil. Initially these phosphate compounds are fairly soluble, meaning they are relatively plant available. Over time, the phosphorus will bind to other compounds in the soil and will change to less plant-available forms.
>
> When phosphorus is bound to other compounds in the soil, very little of it leaches, allowing phosphorus levels to build in the soil.
>
> So now there is more phosphorus in the soil, but most of it is tied up with other compounds. The best way to get that phosphorus into a plant-available form is by promoting soil life and planting a diversity of crops, including cover crops.

year two and year eight, from 8 to 25 ppm. That's a huge jump, but it doesn't reflect anything happening in the soil. I know that sulfur levels on the last soil test are high simply because I pulled the soil sample soon after I'd applied sulphur. On the next soil test, I know sulfur levels will be back down closer to nine ppm again. Because this farm has sandy loam soils and gets over 35 inches of rainfall a year, I don't expect to hold extra sulfur in the soil. If the levels did stay high, it would likely mean there was a hard pan or drainage problem preventing the sulfur from leaching from the soil.

Even though I apply boron to my soils every year in my crop fertilizer, boron levels in the soil have remained the same. Again, this is because boron is an anion and won't build up in the soil. If it rains a lot, sulfur and boron will just wash right out. That's why I apply trace minerals as part of the crop fertilizer so I can feed my crop the traces it needs, and I do not add them as part of a soil corrective since they will leach and not build up in the soil.

Dealing with Excesses in the Soil

What about nutrient excesses? Most consultants can help you deal with deficiencies in your soil, but what do you do if you have too much of a mineral? Correcting the soil is about attaining a sufficiency level in the soil of all nutrients. It is important to apply minerals to make up for deficiencies, and it's equally important to work at bringing down the levels of nutrients in excess in your soil. If one nutrient is in excess, it will impact plant uptake of other nutrients. For example, if potassium levels are too high, they will interfere with the uptake of magnesium and calcium. If phosphorus levels are high, they can interfere with plant uptake of zinc. If calcium levels are too high, it will be difficult for the plant to get enough magnesium. And if boron or copper levels are too high, they become toxic to the plant. Just like mineral deficiencies limit plant health and yield, mineral excess are also problematic and will limit yield.

How do we address excesses? Soil correctives can help deal with many of the problems caused by surplus nutrients. For example, if potassium levels are too high, I would definitely not add any potassium to my crop fertilizer, but I would apply calcium to help kick some of that potassium off the clay/humus complex where it is held. If phosphorus levels are more than ten times the level of zinc in the soil, plants won't be able to access sufficient zinc. In that case, I would back off on any phosphorus application, and add extra zinc sulfate to the crop fertilizer. What about excess calcium in the soil? Certainly don't apply any more lime! I have seen farms with a base saturation of calcium at 80 percent or more that continue to spread high calcium lime each year. The first thing I do when I visit a farm like that is advise them to stop applying lime. When there is that much calcium in your soil, don't waste your money buying more! Instead, apply more potassium and magnesium to ensure that plants are getting enough of those key nutrients.

The Financial Side

Some soil consultants will recommend fixing your soils based on your soil test as quickly as possible. If it has been a while since any soil correctives were applied and your soils are out of balance, that could be an extremely expensive proposition. If a farmer is handed a recommendation that is three or four times higher than his fertilizer budget, what do you think he will do? Many would get discouraged and do nothing. When planning your soil correctives, it is more important to stay within your budget and keep working toward that goal of balanced soils than it is to fix everything right now.

The past few years have been difficult for farmers. Fuel, seed and fertilizer prices have skyrocketed, but crop prices have not kept pace. What do you do when the economy is poor and your inputs budget is low? Your soil has a certain ability to dish out nutrients, and based on soil tests you can formulate a minimal crop fertilizer to get through a lean year. But keep in mind that by doing this you are mining your soil, robbing from the soil's store of nutrients. Eventually you will have to replace what you have used and restore those depleted reserves; otherwise your yields and crop health will suffer.

What about farming on rented ground? A lot of farmers, like me, own a home farm and rent other nearby land. There are a lot of rented acres out there, and farmers are reluctant to spend money to fix the soils on land they might not be farming next year. I can't blame them. Why spend money to fix land that isn't yours and you may not get to reap the benefits? I rent a lot of acres myself, and I understand the dilemma. On the one hand, it is important to me that my soils are balanced so that I can reduce yield-limiting factors and grow a healthy crop. On the other hand, I don't want to spend a lot of money for soil correctives when some of those benefits won't be seen for years to come — when it is possible someone else will be farming the land.

I always test my soils, whether I own the land or rent it. I want to know the soil's ability to dish out nutrients and I want to know what nutrients are lacking or in excess. If the soil on the rented land is in poor shape, I will apply 1,000 pounds per acre of chicken pellets each year in addition to starter fertilizer. The chicken pellets are a crop fertilizer with a good source of nitrogen and phosphorus, but they also have some elements of a soil corrective. Not all of the minerals from chicken pellets will be plant available the year I apply them, but those not available this year will stay in the soil and slowly become available over time. In that way, the chicken pellets act like a soil corrective. It may seem like I'm spending money on fertilizer that I'm not getting back, but I will see the payback down the road as soil biology is stimulated and more nutrients become available to my crop. Also, having a local manure source makes my decision to rent run-down land easier. When I have a good source of inexpensive nutrients nearby, I know I can successfully improve the fertility of rented land without spending too much.

I recently pulled some land out of Conservation Reserve Program (CRP) to put into hay. It is a commonly held myth that putting land into CRP improves soil fertility. That former CRP land didn't even grow good weeds when I worked it up! The first year I applied a ton of chicken manure per acre to kick start fertility and get things going again. It is rented land, so you may think I am wasting my money, but I know I will be farming that land for at least the next five years and I will see a return on my investment. Without any correctives, that soil wasn't going to grow much of anything. It is not worth my time and the cost of seed to plant a field that is depleted in minerals that I know won't grow a good crop.

The Role of Cover Crops as Soil Correctives

The other aspect of correcting soils is improving soil biology. Soil life plays an essential role in holding nutrients in the soil and in converting them from forms plants can't use to plant-available forms. On my farm, building soil biology is an essential step in soil correction, and cover crops are a key part of that program.

On the Otter Creek Organic Farm, the first year I owned the farm I planted corn interseeded with clover, hairy vetch and rye grass. The interseeded plants provided me with plant diversity to feed a diversity of soil life (something I'll discuss further in a later chapter), and additional plant material to work back into the soil in fall. I also planted winter rye in the fall as an overwinter cover crop. I never miss an opportunity to have something growing to feed soil life and cycle nutrients. Without soil biology it is difficult to get enough of all of the needed nutrients into a plant, so for me, building soil biology is central to my soil correctives program.

Another strategy I like to use is planting an oat cover crop after I've applied chicken pellets. If I apply chicken pellets and then plant oats, as the oats grow they suck up all the soluble nutrients they can get. By doing this, not only have I robbed nutrients from the weeds that are going to be coming, I've now stopped those nutrients from leaching, eroding, or tying up with something else. The nutrients are captured in the oat plant, and when I work the oats back into the ground, they will release most of the nutrients they hold back into the soil for the next growing crop.

Just because you have a "perfect" soil test (meaning it is very close to the recommended levels for all of the minerals tested and pH) doesn't mean you can or should do nothing to improve your soils. Soil biology plays a large role in making nutrients available to plants, so no matter what your soil test looks like you need to be farming to promote soil life. Remember, you need to address all three aspects of soil health: chemical (soil nutrients), physical (soil structure and tillage) and biological (soil life and cover crops).

How Long Does it Take to Fix the Soil?

Another question farmers often ask is, "How long does it take to fix the soils? If I add rock phosphate, high calcium lime and Bio-Cal and I grow my green manure crops, what can I expect to see 3-5 years later?"

This is a difficult question to answer because it varies so much from farm to farm. Soil type, crop rotation, crops grown, farm history, weather, and the types and amounts of nutrients applied all play a role. On our main dairy farm, in this example, I added over a ton of calcium and half a ton of phosphorus over the first two years, and saw big increases in the levels of both of these nutrients in the soil after just two years. But this type of response won't happen on every farm in every situation. Some farmers see a great crop and soil response after just one year of applying soil correctives and growing green manure crops, while on other farms it takes longer. If you don't see a quick response on your farm, it is important not to get discouraged and give up. It can take time to fix the soil, but once it is fixed, you will reap the reward of higher yields and healthier crops. In the meantime, while working to improve your soils, you can address nutrient deficiencies in your crop fertilizer.

My goal in applying soil correctives and growing green manure crops is to get my soils active enough that the need for soluble supplementation (the crop fertilizer) goes down. For example, if my phosphorus levels are low in the beginning, I might spoon-feed a little more soluble phosphorus

The Three Components of Soil Health

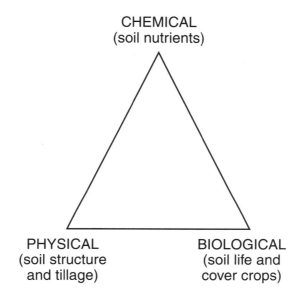

CHEMICAL
(soil nutrients)

PHYSICAL
(soil structure
and tillage)

BIOLOGICAL
(soil life and
cover crops)

in my fertilizer until the soil levels are high enough that I can back off on the fertilizers. The same is true for calcium, potassium and magnesium levels in the soil. Once you have put the time and money into correcting the soil, it doesn't mean your job is done. Even with a great looking soil test and abundant soil life, you still need to feed the crop each year. And since some of the trace minerals are anions, they do not build up in the soil and need to be applied every year. Soils with a good balance of nutrients have fewer limiting factors on yields and crop health, so once soils are corrected, fertilizer inputs can go way down while yields will stay the same or go up.

Crop Fertilizer

The crop fertilizer is how you maximize your production potential. Your soil has a certain ability to dish out nutrients, but by adding a crop fertilizer you provide more of the nutrients your crop needs this year, and in a more readily available form. You won't reach your maximum yield potential without adding a crop fertilizer — especially if you're farming lighter soils.

I want all of my crop fertilizers to contain a balance of nutrients, including trace minerals. My motto for coming up with a crop fertilizer is "balance, concentration, recovery." I want to supply my crop with a balanced nutrient source that is a blend of soluble and slow-release fertilizer sources so it is not overly concentrated at the beginning of the season, and so the plants can recover those nutrients throughout the growing season.

For the forage crop fertilizer program on my farm, I apply homogenized trace minerals in the springtime and I put two tons of compost on every acre in the summer. I don't want to apply too many nutrients at the beginning of the growing season or I'll end up feeding a healthy crop of weeds. I want to get the traces on right away at the start of the growing season so my crop isn't short, but I wait until my crop is up and taking up more nutrients before applying compost.

Following the conventional NPK (nitrogen, phosphorus, potassium) model of fertilizing is about feeding the crop just enough to get yields each year. But what about all of the other minerals? What about calcium, sulfur, zinc and boron? The crop still requires these and other nutrients not found in an NPK fertilizer. Over time, this type of farming mines the soil of all of its secondary and trace nutrients, to the point where eventually they will become limiting factors on yields, if they aren't already.

Soil Type & Crop Fertilizer

Some soils don't dish out nutrients as easily as others. Those with less clay and organic matter, like sandy soils, have a lower CEC which means they have less ability to hold on to nutrients. On soils like this, soil correctives won't bring the levels of nutrients up very high; I still have to spoon-feed crop fertilizers in order to get good yields. Some of the land I farm is very sandy, and even though the soil doesn't retain nutrients very well I can still grow 200 bushel corn. I do apply soil correctives, but with the understanding that they aren't going to move the level of nutrients as far as they would on heavier soils. I also try to build organic matter in the soil by growing cover crops and applying manure, compost or humates. To grow a good crop on sand requires more work and more crop fertilizer than farming heavier ground. You need to be out there all the time, paying attention and monitoring, because you don't have a large nutrient reserve and you don't have organic matter to hold on to the nutrients. In general, sandy soils won't yield as much as heavier soils that have a much greater ability to hold on to nutrients and water.

Using a soil test to determine how much soil corrective and how much crop fertilizer to apply to balance your soils and grow a healthy crop is a little bit like playing football. I take soil samples and tissue tests to get an idea of what is happening in my soil, but it is just an estimate. If I pull two tissue tests and one soil sample from a 30-acre hay field, the handful of plant tissue sent to the lab has to represent a 60-ton hay crop, and the two cups of soil I sent to the lab has to represent 120 million pounds of topsoil. So obviously the test I get back from the lab is not going to tell me exactly what is going on in all of the plants on that field or in all of the soil on those 30 acres. From that soil and tissue test though I need to determine exactly how many pounds per acre of soil corrective I'll add, and how many pounds per acre of fertilizer and micronutrients I'll put in my crop fertilizer.

Soil and tissue test results are a little bit like how we determine a first down in football (if you're not familiar with American football, the ball has to move forward 10 yards to get a "first down" in four tries or less. If you make those ten yards, you get four more tries to move the ball another 10 yards. So successfully getting a "first down" allows your team to keep the football and keep moving toward a score.). The quarterback hands the football to the running back and he takes off running through the defenders. At some point he gets tackled and a bunch of guys pile on top of him. He reaches as far forward with the football as he can, and then the referee pulls the pile of people off of him and places the football where he thinks it probably stopped moving forward. Then the referees on the sideline come out with their chains and measure to a quarter of an inch whether or not the football went far enough forward to be a first down.

In both of these situations, on the soil test and on the football field, we're making an exact measurement based on a rough estimate of where things stand. There are a lot of variables and a lot of uncertainties that go into getting the soil test and the tissue test, yet we're going to formulate our inputs down to the pound based on these numbers. But the tests aren't an exact science. Testing gives us clues as to what is happening in the soil, and tells us whether or not we are making forward progress on getting the soil fixed and addressing limiting factors. There is more than one road to Rome, and there is more than one way to fix your soil. The important thing to do is always to be working toward improving soil health through soil correctives, green manure crops, and proper aeration, while monitoring with soil tests and tissue tests to find out if your corrective programs are working.

It is also important to separate soil correctives from crop fertilizers. Soil correctives are a long-term investment in your soil. They are the minerals added to the soil in order to balance nutrient levels so all nutrients have a baseline level of sufficiency, and no nutrients are in excess. Crop fertilizers are a short-term additive. They are the food your crop will consume this year. Crop fertilizers are very important for producing a healthy, high-yielding crop, but their purpose is not to stick around in the soil for years to come or to balance soils. Their purpose is to feed the crop that is growing on the field right now. It is important to separate these two types of inputs and include both in the nutrient program on your farm. Each has its place when working toward your goal of healthy soils and healthy, high-yielding crops.

Chapter 8

Fertilizer: What, Where & When

Rule number two of the Six Rules of Biological Farming is about fertilizer. *Use fertilizers that do the least damage to soil life and plant roots. Watch salt and ammonia levels. Use fertilizers with a balance of nutrients, a balance of soluble to slow-release ingredients and a controlled pH. Use homogenized micronutrients, add carbon and place all properly to enhance performance.*

One of the fundamentals of biological farming is knowing what fertilizer fits your farm and works the best to feed your crop without damaging soil life. Another important aspect to managing fertilizer on your farm is knowing where to place fertilizer to get the most out of it, when and how to apply it, and how best to spend your fertilizer dollars.

Fertilizers today are sold based on price and solubility. How well the fertilizer performs, whether or not there are any environmental issues, and whether there are any negative impacts on soil life are not addressed. If you buy a bag of fertilizer, you will see three numbers on the tag, 9-23-30 for example. Those numbers mean the fertilizer in the bag contains nine percent soluble nitrogen, 23 percent soluble phosphorus (in the form of P_2O_5), and 30 percent soluble potassium (in the form of K_2O). The higher the soluble fertilizer content, the bigger the numbers on the fertilizer tag. My grandfather used to buy a 3-12-12 fertilizer. How was his 3-12-12 different from my 9-23-30 fertilizer? Both bags were full! The lower numbers on the tag mean the fertilizer my grandfather bought was less soluble. Over time fertilizers have been purified, and now most commercial fertilizers consist of purified water soluble nitrogen, phosphorus and potassium. In my opinion my grandfather had a better balanced fertilizer than we have today because a lot of the other minerals found in fertilizer were lost when we started chemically treating mined minerals to purify them. Purifying mined minerals, i.e. breaking them down and keeping only the wanted nitrogen, phosphorus and potassium (NPK), removes nutrients like calcium, sulfur and trace minerals. I recommend fertilizers that contain naturally mined minerals and a balance of soluble to slow-release sources, which means they have lower tag numbers. But the lower tag numbers do not mean the fertilizer is of a lower quality than a fertilizer with higher tag numbers. In fact, the opposite may be true. Fertilizers that contain "extra goodies," like the trace minerals found in naturally mined minerals, can perform better and may address other limiting factors that a commercial fertilizer will not.

Sometimes you will see a fourth number on the fertilizer tag. For example, one fertilizer I like is 0-0-50-17S, potassium sulfate. This fertilizer contains 50 percent K_2O potassium and 17 percent sulfur. It is interesting to note that only nitrogen, phosphorus, potassium and sulfur appear on the fertilizer tag. If a fertilizer contains any other elements, like calcium, magnesium, or trace elements, you won't find those numbers highlighted on the tag. The only thing the fertilizer tag numbers tell is how much soluble NPK and S are in the bag.

Soluble Fertilizer & Plant Needs

I talk a lot about the problems with soluble fertilizers, but there is a reason why chemical companies make their fertilizers soluble. That's because the nutrients a plant can absorb are those that are found in the soil solution — meaning that plants need soluble nutrients. Getting some soluble nutrients from your fertilizer is not a problem. The problems arise when you put on too much soluble fertilizer all at once, or use a source of fertilizer that is hard on roots and soil life. This is similar to what happens when we eat sugar. Sugar is a good source of energy for our bodies, and getting some sugar from eating fruits and vegetables is good for us. Eating a lot of refined sugar from candy, sweets and soda pop is not healthy. This is too large a dose of sugar and it's in a form that can cause health problems.

Nutrients in the Soil

Nutrients are found in the soil in different forms. Some of the nutrients are tied up as part of the soil matrix or as slowly decomposing plant material, like mature plants and wood. Some of the nutrients are in an intermediate state where they are held on clay particles or humus in the soil. And some of the nutrients are soluble in water, meaning they are in the soil solution. The only nutrients plants can access in the soil are those in the soil solution.

The soil solution is very dynamic. Nutrients are constantly moving in and out of it. Nutrients leave the soil solution when they are taken up by plants or microorganisms, when they interact with other nutrients in the soil to form less plant-available compounds (what we call tying up of nutrients), or when they are leached out of the topsoil. Nutrients can move back into the soil solution from where they are held on humus or clay particles, or be released into the soil solution from decay processes by microorganisms. Soluble nutrients can also be added to the soil solution from inputs like manure or fertilizer.

Another problem with conventional fertilizers is that most farmers apply all of the fertilizer their crop needs for the season in the spring. That is a large dose of soluble nutrients all at once, applied when the crop is not yet using many nutrients. Nutrients need to be continuously available throughout the growing season in order for plants to be healthy. In the spring, when most fertilizers are applied, plants are small and only utilizing a small amount of nutrients. The young plants can't take up that large dose of soluble nutrients, and in fact all of those easily available nutrients in the soil solution may discourage vigorous root growth. Over the course of the growing season, as the plants get larger and become reproductive, they need more nutrients. But the amount of applied soluble nutrients in the soil declines over time, leaving less available when the plant needs them the most.

The following chart shows how soluble phosphorus levels from a conventional fertilizer begin to decline only 30 days after application. In contrast, soluble phosphorus levels in the soil from a balanced fertilizer source remain higher throughout the growing season. Compare that to plant phosphorus needs, which really take off about 35 days after planting.

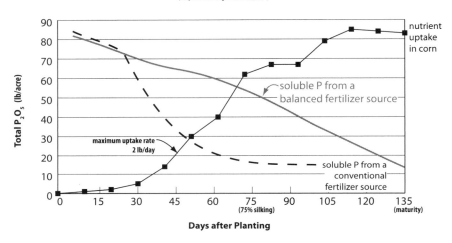

Phosphorus Sources and Uptake in Corn
(adapted from Mengel and Barber, 1974)

Where do the extra soluble nutrients go? Because the soil solution is dynamic, excess nutrients don't just hang out in their soluble form in the soil solution across the whole growing season, waiting to get taken up by plants. Some of them will leach out of the soil, some will volatilize into the atmosphere, some will be utilized by growing plants and microorganisms, and some will bind up with other compounds in the soil. This means that over time, less and less of the applied soluble nutrients are available to fill the needs of the growing plant.

Later in the growing season, when corn is tall and it is showing signs that it doesn't have enough nutrients, you might think that you didn't apply enough fertilizer in the spring to cover the plant's needs. Your solution might be to apply more soluble fertilizer next spring so you can be sure the plant won't run out. But putting on more fertilizer in the spring to address a shortage later in the growing season isn't going to solve your problem. You need nutrients to be available to your crop throughout the growing season. One option is to split a fertilizer application, providing some in spring and more later in the growing season. Another option, one that I use on my farm, is to apply a balance of soluble and slow-release minerals in my starter fertilizer. That way nutrients are available up front when the plant is starting out, but there are also available nutrients later in the growing season when the plant needs large amounts of nutrients each day for flowering and seed production. It is also important not to discount the role healthy soils play in making nutrients available. A well-balanced, biologically active soil is going to provide available nutrients to growing plants, especially later in the growing season when the soil has warmed up and the soil life is abundant and thriving.

Soluble Nutrients & Environmental Issues

Running short on nutrients later in the growing season is not the only problem that comes from applying a big dose of soluble nutrients in the spring. Another problem is runoff. What happens when you apply a water-soluble fertilizer and then it rains? Where does all that fertilizer go? It goes into the streams, down the rivers, and down into the water table.

Potato farmers are especially concerned about water pollution. On many potato farms, the soil phosphorus levels are six times higher than what the university recommends, but without adding additional soluble phosphorus, the farmers get half the yields. Since they get such a big response from applying phosphorus, each year they put on around 150 lbs. per acre. The problem is that much of it runs off and gets into our waterways causing algal blooms. Many states have responded by regulating phosphorus applications. In my home state of Wisconsin, potato farmers have been lobbying against such input regulations. Recently I was visiting a conventional potato farmer and we were discussing the proposed laws regulating fertilizer use. I noticed that he was drinking bottled water and I told him, "If you're going to lobby against regulating phosphorus and other nutrient inputs, get the bottled water out of your office! That's a dead giveaway that something isn't working right." If the farmer isn't comfortable drinking the water from his own farm, doesn't that indicate that there's a problem with applying too much soluble fertilizer and other chemicals?

Soluble Nutrients and Plant Disease

On a farm tour a couple years ago I met a university professor who went from farm to farm looking at soils and measuring the electrical gradient in the soil using a conductivity meter. (Conductivity is a measure of soluble nutrient levels. The higher the conductivity of the soil, the more soluble nutrients there are.) The professor told me that he had never seen root-eating nematodes on an organic farm. He had no explanation for why that was, other than that when he measured conductivity in the soils on organic farms it was very low.

It makes sense to me that conductivity is low on organic farms because commercial soluble fertilizers can't be used. Organic farms must get their nutrients from naturally mined minerals, manure, compost or green manure crops. These sources are less soluble, so would not cause a high conductivity reading in the soil. Highly soluble commercial fertilizer sources result in a high level of conductivity, which in this professor's experience was related to the presence of root-eating nematodes.

On this same farm tour there were several older farmers who said that a long time ago farmers planted sudangrass before their potato crop so they would not have root-eating nematodes infesting their potatoes. For 20 years they have been relating that story to researchers, and no one has been able to come up with an explanation for why that would be. University staff measured the sudangrass to see if it had any biofumigants in it (chemicals that kill off nematodes and fungi in the soil) but they never found any. In my opinion, the reason you don't find root-eating nematodes in the soil after planting sudangrass is that the plants take a lot of soluble nutrients out of the soil. When the soluble nutrients in the soil are tied up in the sudangrass plant, the nematodes go away. There was really no nematode-deterring chemical in the sudangrass at all — I think it was all related to the level of soluble nutrients in the soil.

Things I Do Not Want in my Fertilizer

High salt index: Salt index is a measure of the salt concentration that a fertilizer induces in the soil solution. Almost all fertilizers are a salt, which simply means that the fertilizer is made up of both a positively charged ion (a cation) and a negatively charged ion (an anion). For example, when you want to apply calcium to your soil you don't buy pure calcium, you buy calcium sulfate, calcium oxide, calcium carbonate, calcium nitrate or another compound where calcium is hooked to something else. Calcium has a positive charge, so whichever source of calcium you use, it will always be hooked to some other substance with a negative charge. The same is true of almost all fertilizers. The salt index is a measurement that tells you how strongly the cations and anions are hooked together, and how quickly they break apart in solution. Compounds that move quickly into solution have a higher salt index.

Fertilizer Material	Salt Index
Potassium chloride	116 ✗ Don't use
Ammonium nitrate	105
Sodium nitrate	100
Urea	75
Ammonium sulfate	69
Calcium nitrate	53
Potassium sulfate	46
MAP	30
Calcium sulfate (gypsum)	8
Sodium chloride (table salt)	154

It is important to know the salt index of the fertilizer you buy because high salt index materials can injure germinating seeds, or damage roots and soil life. When a salt fertilizer goes into solution, it can cause plant roots to lose water and dry out, which can kill the roots and stunt plant growth. The addition of a large dose of a high salt index fertilizer can also harm or kill soil life.

Unwanted compounds: When buying fertilizer, it is important to consider not only the compound you want, but also the compound it is hooked to. For example, potassium chloride is a popular commercial fertilizer made up of positively charged potassium and negatively charged chloride that supplies a readily available source of potassium to crops. However, potassium chloride is 47 percent chloride. When potassium chloride is applied to the soil and goes into solution, the chloride portion can cause damage to chloride sensitive plants and to soil life. Plants do need a small amount of chloride as a nutrient for growth, but not nearly the amount supplied by adding potassium chloride fertilizer to meet potassium needs.

In addition, applying potassium chloride can result in a loss of soluble calcium from the soil. When the negatively charged chloride ion goes into solution, it will bind to positively charged compounds in the soil. Calcium is a positively charged compound that combines readily with chloride to form calcium chloride. The formation of calcium chloride actually removes calcium from the plant-available pool, which can cause a big reduction in the amount of available calcium in the soil. Available calcium is needed for early root development and growth, which means that if a farmer applies potassium chloride in the spring and ties up available calcium, there may be problems with plant growth and health later on.

Compounds that reduce soil carbon: Anhydrous ammonia is a highly concentrated source of ammonia. When anhydrous is injected into the soil, the high concentration of nitrogen stimulates soil bacteria. Soil bacteria consume carbon as well as nitrogen, and as their activity increases they release large amounts of carbon from the soil into the atmosphere as carbon dioxide. Anhydrous also solubilizes humus in the soil. Humus is a stable carbon that provides a multitude of benefits to growing crops. Eventually, this loss in soil carbon and humus can lead to changes in soil structure, making soil hard and tight.

Too much soluble nitrogen: Overuse of soluble nitrogen from any source takes a toll on soils over time. A study conducted by Dr. Phillip Barak at the University of Wisconsin-Madison found that applying too much soluble nitrogen interferes with the soil's ability to hold on to calcium, magnesium and potassium. He estimates that the damage to the soil, done primarily over the last 30 years of agriculture, has aged our Wisconsin soils 5,000 years. In addition, if farmers are putting on soluble nitrogen and not watching pH or applying any lime, over time the soil pH gets lower and lower. Soil life struggles at the lower pH, the soil can't hold on to as many nutrients, and crop health and yields suffer. This alone is a good reason to watch nitrogen inputs and use as little commercial nitrogen as possible.

Farmers can reduce the need for soluble nitrogen by growing their own nitrogen through green manure crops and healthy soils. In addition, adding carbon (from compost or humates) and sulfur to nitrogen blends can reduce some of the negative effects, as can placing nitrogen when and where it is needed.

Things I Do Want in my Fertilizer

A good fertilizer should enhance the natural processes taking place in the soil. When you use a biology friendly fertilizer, not only do plants get the nutrients from the fertilizer you applied, they also get nutrients provided by the increased activity of soil organisms. Things I like to have in my fertilizer include:

- Quality nutrients that work with soil life
- Naturally mined minerals
- A blend of soluble and slow-release sources
- Acidity
- Carbon, as a buffer
- A balance of all nutrients, not just NPK

Quality nutrients that work with soil life: I am less concerned about the numbers on the fertilizer tag than I am about the quality of the ingredients that make up the fertilizer. Fertilizer ingredients should promote soil life as well as provide soluble and slow-release nutrients to my crops. I want my fertilizer to have a balance of nutrients and to consist of high-quality materials that are not harmful to soil life. Fertilizer should enhance what goes on in the soil, not inhibit it.

Naturally mined minerals: One of the reasons I like to use naturally mined materials is that they contain all of the minerals found in the mined rock, not just the NPK. Crop consultants balance the soils for up to 12 minerals, and yet we know that there are close to 70 minerals needed by plants. Naturally mined minerals contain some of those other minerals that are not measured on the soil test but are needed in small amounts by growing plants. These natural materials are also less soluble than commercial fertilizers, so they stick around in the soil longer and do not leach.

A blend of soluble and slow-release sources: Including naturally mined minerals in my fertilizer blend also helps me balance the ratio of soluble to slow-release nutrients. I want my fertilizer to have some nutrients available right away when I apply it, as well as some that are available later on. If my fertilizer consists only of quickly available soluble nutrients, I will have more nutrients than my crop can use early in the growing season, and I will run short of minerals later in the growing season. On the other hand, if I use only slow-release sources, my crop won't get the boost it needs to get going at the start of the growing season. A blend of the two gives me the balance I want between soluble and slow-release, keeping nutrients available for my crop throughout the entire growing season. One method of blending soluble to slow-release nutrients is by mixing a commercial phosphorus source like MAP with a mined rock phosphate. Another method is to finely grind the mined minerals and mix them with an acid carbon source, such as humates.

Acidity: I like my fertilizer to be slightly acidic. This helps to break down minerals like rock phosphate and limestone that are otherwise slow to move into a plant-available form. Keep in mind that I want my *fertilizer* to be acidic, not my *soils*. Acidic soil is not ideal for soil life, and I do not want to do anything that will inhibit soil biology. Also, if you have a low pH soil it means you are missing minerals, and I don't want to give up my opportunity for a better crop. A moderately acidic fertilizer on a neutral pH soil gives me the balance I want between healthy soil life and plant-available nutrients. Applying a low pH starter fertilizer should not bring down my soil pH. However, I do watch my soil pH closely and apply lime whenever a soil test shows the pH is below 6.8.

Carbon, as a buffer: I also like to buffer my fertilizer with a carbon source. Mixing soluble nutrients with a carbon source like humates or compost gives the minerals something to hook on to so they don't all move into solution right away. Keeping the minerals loosely hooked to a carbon source means I won't have problems with leaching, yet those minerals can easily move into the soil solution in a plant-available form. These attributes of humates and compost greatly increase the efficiency of my fertilizer.

A balance of all nutrients, not just NPK: I think we made a big mistake when we named our "major," "secondary" and "trace minerals."

Every mineral is needed. Calling some minerals "major" and others "trace" attached an importance to them that had nothing to do with which minerals a plant needs. Major, secondary and trace are terms that signify quantity, not importance. Farmers need more of the major minerals, less of the secondary minerals, and just a little bit of the trace minerals. But all of the minerals need to be included in a balanced crop fertilizer, because they all are needed to grow healthy crops.

A shortage of trace minerals will cause crop problems the same way that missing major minerals does. For example, a crop farmer I know in Canada decided to expand so he bought the neighboring farm where hogs had been raised for many years. He tore down the fence between the two properties and started farming crossways from his old farm onto the new farm. His land is along Lake Michigan, and there are a lot of mold problems in that area caused by the high humidity present along the lake. I visited his place at harvest time and was looking around and noticed that the soybeans on his side of the farm had white mold all over them, but the soybeans on the former hog farm side were taller, completely canopied, and had no white mold at all. How could this be? The farmer took soil tests and saw that the copper level on the newly purchased land where hog manure had been applied for many years was over six ppm, while the copper level on his home farm was below two ppm. Years ago, hog manure was high in copper because hogs were fed copper as a drug. In my opinion, the lack of mold on the former hog farm half of the field was directly related to the high copper levels in the soil — a trace mineral that is often overlooked in soil fertility programs.

Some Specifics on Fertilizers

There are a lot of different fertilizer sources out there — some that work well in a biological farming system and others that do not. Knowing the benefits and drawbacks of different fertilizer sources can help you make the right choice for your crop and for your farm. The following table and discussion covers the most common fertilizer sources and should help answer your questions on the pros and cons of using each.

Fertilizer Sources

SOLUBLE ───────────────────────────► SLOW RELEASE

	SOLUBLE		SLOW RELEASE
Manure and green manure sources	Fresh animal manure Young green plants		Compost Mature plants, crop residue
Nitrogen sources	Anhydrous ammonia Ammonium nitrate Urea	Ammonium sulfate Pelletized chicken manure	Legumes (i.e. alfalfa, clover, etc.) Fish meal Feather meal
Phosphorus sources	MAP, DAP, Orthophosphate Polyphosphate		Rock phosphate
Calcium sources	Bio-Cal, OrganiCal HumaCal Calcium nitrate Calcium chloride	Gypsum Calcium sulfate	Calcium carbonate (high calcium lime) Calcium magnesium carbonate, (dolomitic lime)
Potassium sources	Potassium chloride	Potassium sulfate	K-mag Granite dust
Sulfur sources	Sulfate sulfurs		Elemental sulfur
Micronutrient sources	Homogenized sulfate trace minerals (i.e. zinc sulfate, manganese sulfate, etc.)	O-charger Triple five	Chelated trace minerals Oxy trace minerals

Manure

Manure and compost are excellent sources of nutrients because they provide a blend of minerals in a form that is tied to biology. The difference between these sources of nutrients is how quickly they become plant available. Compost is a slow-release source of nutrients while manure is soluble, meaning it is quickly available to plants.

A farmer I work with found out the downside of applying a lot of soluble nutrients in the spring when he went out and applied 8,000 gallons of liquid manure to his fields and then planted soybeans. He was overrun with weeds. Some people say if you don't compost manure the weed seeds in the manure will germinate and cause problems, but in my opinion that is only a small part of it. The bigger problem is the soluble nutrients in liquid manure that cause weed seeds already in the soil to germinate. I have problems with weeds on my farm when I put raw manure on the land in the spring, but I see fewer weeds after I apply compost. I don't believe this happens because I have a problem with weed seeds in my livestock manure. Rather, raw livestock manure is full of soluble nutrients, which sets up conditions for weeds to germinate and grow. The nutrients in compost are stabilized and less soluble so fewer weeds pop up right after applying compost than after applying raw manure.

Green manure crops (or cover crops) and crop residues are also excellent sources of nutrients. Not only are green manure crops a means for holding onto nutrients so they can't leach, tie up or erode, as the plants decompose they feed soil life, which releases nutrients in a very plant-available form. Similar to the comparison made of manure and compost in the previous paragraph, green manure crops and young, succulent plants are a source of soluble nutrients, while mature plants and crop residues are slow release. A young green manure crop worked back into the ground breaks down right away and immediately releases nutrients into the soil. You won't find a trace of that green manure crop two weeks after it is worked into the soil. In contrast I can go out to a field and find corn stalks two years after a corn crop was harvested and the stubble worked into the ground. Mature plant residues break down much more slowly, and the nutrients in them take a long time to become plant available.

Nitrogen Sources

Like other fertilizers, nitrogen is sold based on solubility. If you look at the Fertilizer Sources table, you will see that anhydrous ammonia, ammonium nitrate and urea are all very soluble sources of nitrogen. As I've already discussed, I don't recommend using anhydrous ammonia because I don't believe it fits on a biological farm where the goals are to increase soil organic matter and humus over time.

I don't usually recommend urea because it is unstable and can release ammonia gas into the soil, which is toxic to roots and soil life. Applied urea also needs to be kept at least six inches away from the seed so it does not inhibit root growth. However, using a small amount of urea is not always a problem. It is often found in small quantities in foliar sprays, and in that form I think it works well.

My preferred nitrogen sources are ammonium sulfate and pelletized chicken manure. They both have some soluble and some slow release aspects to them.

I can no longer use ammonium sulfate on my farm because it is certified organic and ammonium sulfate is not allowed by the organic rules, but I would use it if I could. It is excellent for spring application on corn, small grains and alfalfa because it has a warming effect on the soil, which extends the growing season. The other thing I like about ammonium sulfate is that the nitrogen source (ammonium) is hooked to sulfur, which is a needed element in a fertilizer program.

Chicken pellets from laying hens are high in nitrogen (from five to eight percent), and provide nitrogen in a form that is easily digestible by soil microorganisms. Next to manure and cover crops, chicken pellets are the source of nitrogen I use most on my farm.

If more nitrogen is needed on a biological farm, I often recommend ammonium nitrate (liquid 28 percent). Since we want to use as little nitrogen as possible to get the job done, placement, timing and add-ons like thiosulfate, humates or molasses can improve efficiency and allow a reduction in quantity. Another good option for an efficient nitrogen source that can save future trips over the field is polymer-coated urea, labeled as ESN (Environmentally Smart Nitrogen). The nitrogen in ESN is coated in a substance that breaks down from moisture and temperature, slowly releasing nitrogen into the soil. Farmers I know who use ESN have been very satisfied with its performance.

Fish meal, feather meal and animal byproduct fertilizers are also excellent sources of nitrogen, but they are very expensive. They work well as a supplement to other nitrogen sources, but are usually not practical as the sole source of nitrogen for a crop.

Legume cover crops are another excellent source of nitrogen, working well either as a stand-alone cover crop, or when interseeded into other crops. On my farm, I have had good success interseeding clover into my corn crop. Some legumes, like alfalfa and clover, can provide up to 200 pounds per acre of nitrogen per year. A legume cover crop will provide nitrogen two ways: first, as it is growing and fixing nitrogen in its root nodules, and second, when it is worked back into the soil and becomes food for microbes. Also don't forget that cover crops have more benefits than just supplying nitrogen; they also build soil structure, prevent erosion and feed soil organisms.

Phosphorus Sources

Orthophosphoric acid, or orthophos, is a liquid phosphorus source used as an ingredient in many high quality liquid fertilizers. It is a readily available source of phosphorus for plants. However, because of its chemical make-up, it ties up quickly with other elements in the soil and may become unavailable within hours of application. Polyphosphoric acid, or polyphos is produced by dehydrating orthophos. This process makes it more stable so it stays in the soil longer before tying up with other elements.

MAP and DAP (monoammonium phosphate and diammonium phosphate) are highly soluble dry phosphate fertilizers. Both also contain nitrogen in the ammonium form. MAP has a lower pH and less ammonium than DAP, making it a better source of soluble phosphate and is easier on soil life. The commercial fertilizer industry makes soluble phosphorus fertilizers like MAP and DAP by taking insoluble rock phosphate and mixing it with an acid, like sulfuric acid, to create orthophosphoric acid. The phosphorus is then purified out, which means calcium, sulfur and other beneficial elements found in the rock phosphate are removed. The final step is to mix the purified orthophosphoric acid with ammonia to create

MAP (monoammonium phosphate) or DAP (diammonium phosphate). This process makes a highly soluble phosphorus source, but all of the other elements of the rock phosphate have been removed.

I generally do not recommend DAP. It has a high pH which can damage root hairs, those fine hairs on roots that take up most of the water and nutrients plants consume. DAP is also high in ammonia and can release ammonia gas into the soil, which is hard on soil life.

My preferred phosphorus source is a blend of rock phosphate and a commercial soluble phosphorus source such as MAP. I like to include rock phosphate in the blend because I want to keep the calcium, sulfur and trace elements found in the naturally mined rock. I also don't want to overdo application of soluble nutrients, which in the case of phosphorus ends up being a waste of my money since much of the phosphorus from a soluble source will tie up quickly in the soil. By applying a mix of rock phosphate and commercial phosphorus, I get a good blend of soluble and slow-release phosphorus.

If I have acidic soil that needs phosphorus and calcium, that is the perfect time to add a rock phosphate soil corrective. The acidity in the soil will speed up the breakdown of the rock phosphate, and I get as much calcium out of it as I would if I put lime on. If I don't have an acidic soil, it will take a long time for the phosphorus to become plant available unless I have abundant soil biology. Regardless of soil pH, phosphorus uptake is tied to soil biology. Planting a cover crop like oats, rye or buckwheat can stimulate soil biology and help plants access phosphorus in the soil. Plants with more acidic roots, like oats and buckwheat, can extract more phosphorus from the soil reserve and from rock phosphate. These plants hold that phosphorus in their tissues, putting the nutrient into a biological cycle. This interaction is a vital part of the system. If you put rock phosphate on a hard, dead soil without any life in it and no green plants growing, the opportunity for that phosphate to show up is pretty minimal.

Calcium Sources

High calcium lime (close to 35 percent calcium) and dolomitic limestone (close to 20 percent calcium and 12 percent magnesium) are mined calcium sources that are very slow release. They are a good source of calcium for acidic soils. Just as acidity helps release the nutrients from rock phosphate, acidity breaks down high calcium lime or dolomitic lime. On soils that are neutral or higher pH, these sources will not supply much plant-available calcium. To get more calcium on soils that are not acidic, a source of calcium that's more soluble is needed.

When I started working as a farming consultant I went in search of a soluble calcium source. I found a source of lime (calcium carbonate) that was finely ground, had been burnt in a kiln, and then hydrated to remove the caustic effect of burnt lime. At the time I had no idea that by putting calcium carbonate through a kiln the carbon was burned off and what was left was soluble calcium. In addition, being a natural, mined material and a byproduct of manufacturing meant this calcium source also had some sulfur and other beneficial materials in it. When I applied the hydrated burnt lime to the ground, I got a calcium response in the plant right away. It worked wonders on my alfalfa crops. Later my partners and I developed a product from the hydrated burnt lime called Bio-Cal. Over the years I've seen wonderful responses from the application of Bio-Cal, especially on legumes. Unfortunately, I can no longer use Bio-Cal on my organic farm because it is burned and thus it is considered synthetic. We therefore developed OrganiCal to take the place of Bio-Cal on organic farms. It is a soluble source of calcium similar to Bio-Cal, but rather than burning the limestone it is finely ground and blended with acid binders and sulfur. This makes it more plant available than straight limestone, and because it is not burned or processed, it is approved for use on organic farms.

HumaCal is another calcium product my colleagues and I developed. It is a blend of finely ground limestone and gypsum with humates. Humates are large, complex molecules that have a low pH and contain a lot of sites that hold on to nutrients like calcium. This means that humates can help break down rocks like limestone into a plant-available form, and can also hold on to the plant-available nutrients so they don't leach or tie up. This makes humates an excellent material for blending with a lot of nutrients, including calcium. I have done quite a bit of research on HumaCal demonstrating that it provides plant-available calcium, and I'll talk more about this in the next chapter.

Gypsum, which is calcium sulfate, is more soluble than lime. I like to use gypsum on my land when the soil is high in magnesium because gypsum is not only a source of calcium, it also supplies sulfur. The sulfur will hook to magnesium in the soil and form Epsom salts (magnesium sulfate) which is very soluble. That means it makes the magnesium more plant available but it also leaches, so it washes some of the excess magnesium out of the soil.

Calcium nitrate and calcium chloride are both very soluble sources of calcium. Calcium nitrate is often used as a foliar on high value crops because not only is it a good source of available calcium, it also supplies soluble nitrogen. However, it is a very expensive way to provide calcium, so it is generally only used on high value crops like potatoes and other vegetables. Calcium chloride is better known as road salt. It is also used as a foliar spray, but less often. Even though it supplies soluble calcium, it does have chloride, which has some negative side effects.

Potassium Sources

Potassium chloride is a highly soluble source of potassium, but it has a lot of negatives. Whenever I buy a fertilizer I want to know exactly what is in it — both the element I want and whatever substances it is hooked to. In the case of potassium chloride, the potassium that I want is hooked to chloride, and I don't need very much chloride at all. Potassium chloride is a strong salt, and the chloride in it is hard on roots and soil life. For these reasons, potassium chloride does not fit well in a biological farming system. A small amount of it shouldn't hurt, but placement and timing are critical.

On my farm I use potassium sulfate and K-Mag. K-Mag, (potassium magnesium sulfate), is very slow release, and potassium sulfate is just a bit more soluble than K-Mag. K-Mag is an especially good potassium source on soils that are also lacking in magnesium. On my high magnesium soils, most of my potassium comes from potassium sulfate. Both of these, potassium sulfate and K-Mag, provide two desired elements: potassium and sulfur. Both fertilizers also have potassium in a desirable plant-available form that does not harm roots or soil life, and they can be applied at lower application rates.

Granite dust and green sand are natural mined minerals that contain four to eight percent potassium, plus they contain magnesium, iron and other trace elements. Green sand and granite dust are very slow-release sources of potassium that are most often used by organic farmers and gardeners.

Sulfur Sources

Sulfur is a necessary element for healthy plant growth, protein formation and humus building. It is also an anion, which means it won't build up in the soil, and therefore needs to be applied each year. Sulfate sulfur is plant available because it's relatively soluble, while elemental sulfur is considerably less soluble. Elemental sulfur is not usable by plants unless soil bacteria convert it into the plant-available sulfate form. Elemental sulfur is also problematic because it's a harsh compound that can be hard on soil life, and it acidifies the soil.

I get most of my sulfur by using the sulfate form of calcium, potassium, and trace minerals (i.e. calcium sulfate, potassium sulfate, copper sulfate, etc.). If my farm was not organic, I would also use small amounts of ammonium sulfate. Animal manure and cover crop residues also contain small amounts of sulfur, and round out my sulfur application program each year.

Micronutrient Sources

Micronutrients, or trace minerals, are found in the "oxy" form (i.e. copper oxide), the "sulfate" form (i.e. copper sulfate), and in chelated form (i.e. copper chelate). Chelated trace minerals and oxy trace minerals are less soluble than sulfate forms of trace minerals. Chelated forms are also more expensive, generally only found in liquid or foliar fertilizers, and applied at low rates for plant use rather than being added to the soil. Oxy form trace minerals are the least expensive to buy, but they are not very plant available. The other downside to the oxy form is that rather than being hooked to a needed element, such as sulfate, they are hooked to oxide (oxygen).

On my farm I like to use homogenized trace minerals and always in the sulfate form. I use a homogenized form for a very good reason. To see this, come next winter if you are really bored and looking for something to do, I have an experiment for you. Go out and buy a pound of pelleted copper sulfate and count out how many pellets are in that pound. Then go out and measure out one acre across your field using one-spreader width. Now calculate how close together those pellets are if you apply one pound per acre of pelleted trace minerals. I haven't counted them myself, but someone told me they would be two feet apart. Until someone goes out and conducts this experiment and tells me the answer is different, I'll stick with that number. So if you go out and plant your corn on this field, the plant that comes up right by a pellet will get an overdose of copper sulfate, the plant two feet over by the next pellet will get an overdose, and all of the plants in between won't get any copper sulfate. The corn by a pellet will turn white when it comes up from too much copper. The problem with this hit or miss approach is why I like to finely grind and homogenize my trace minerals so that they are spread evenly across the field, and no plant gets an overdose while the guy next door gets none. I also like to blend my homogenized trace minerals with a carbon source like mined humates that can hold on to the traces so they do not leach and are more readily available for plant uptake.

The What, How Much, Where and When of Fertilizer Application

Liquid or Dry Fertilizer?

won't help soil - only the crop

Liquid fertilizer is fertilizer that is already in solution, which means that the minute you apply it the plant can take it up. This has both advantages and disadvantages. I am not opposed to using liquid fertilizers, but I just don't think you can be very successful if your entire fertility program consists of liquid fertilizer. Liquids have their limitations.

The function of liquid fertilizers is different from that of dry fertilizers. Liquids do not work as soil correctives because they do not stick around in the soil. They also don't work well as the bulk of a fertility program because it is difficult to get adequate potassium or calcium for an entire growing season out of a liquid. You may not be able to get enough of the major minerals from a foliar application, but applying foliar trace minerals can be an efficient means of supplying adequate trace minerals and sulfur. Adding liquid trace minerals and sulfur to a fertilizer or foliar application can deliver a small amount of those minerals for a boost to that crop, but it will not raise the soil sufficiency level. Liquid also costs more than a dry blend. In a typical 200 pound per acre dry corn blend, besides adding nitrogen, phosphorus and potassium I also add calcium, 25 pounds of sulfur, three pounds of zinc, three pounds of manganese, one pound of copper and one pound of actual boron. I also add a carbon source and balance the soluble and timed-released minerals, all for about $75 per acre. It is not possible to do this with a liquid fertilizer program, and certainly not at $75 per acre. A combination of a few gallons of liquid plant food applied in the row at planting and a good balanced dry fertilizer program can supply sufficient minerals and at an affordable price. Foliar spraying is also a good method for adding brewed bugs from compost tea, a small amount of plant food, or a carbon source like liquid humic substances, molasses or sugar.

There is a real movement in some parts of agriculture to apply all liquid fertilizer, not because liquids are a better source of nutrients, but because they are easy to work with. But what happens to liquid fertilizer when it rains? The fertilizer in a liquid is all soluble and it will wash away in a heavy rain. Liquids also tend to be high salt solutions. Another issue with liquid fertilizer is that you can't balance soluble to slow-release sources. All of the fertilizer in a liquid is soluble and available immediately. The problem with this is that crops do not take up all of their nutrients in one dose. My cows need to eat every day, and plants are no different. I want to dish out food for the crop over the entire growing season, which is one reason why I wouldn't recommend using liquids as your entire fertilizer program.

Some farmers I met in Australia have had great success adding a foliar spray to their soybean fertilizer program. During the 20-day period of pod development it is estimated that the soybean plant needs 11 pounds of nitrogen a day. The Australian soybean farmers take urea and humates and mix them together and foliar feed the plants when they are in the early pod stage, adding magnesium as well to boost photosynthesis levels. By giving the plants that extra kick of nutrients in the middle of the growing season they estimate that they get 10 additional bushels of beans per acre. This group is very excited about their success and asked me to pass along their advice. I told them I would, but that I doubted I would find even five farmers to try it. Why not? What happens if it is a drought and the foliar spray does not give you 10 extra bushels of beans? You just went and ran over your beans applying the foliar in the middle of the growing season, and now your neighbors think you're nuts — and you didn't even get more beans for your trouble!

I do not have a problem with this type of foliar application of liquid fertilizers based on plant needs. I am sure that under the right conditions applying a foliar spray of urea and humates to a soybean crop in the middle of the growing season will give a great pay back. However, this is not a replacement for a good fertility program. The biggest problem I have with liquid fertilizers is when they are poured on in the spring as the only source of nutrients for the crop. That is too large a dose of soluble nutrients all at once, and the fertilizer won't be there later in the growing season when the plants need nutrients. It is better to use liquids and foliars as one part of a fertility program that includes a balance of soluble and slow-release, dry fertilizer sources.

How Much Fertilizer to Use

How much fertilizer to apply depends on the condition of your soils, the crop you are growing, and your budget.

The first time I travelled to Western Australia I got a real education on different approaches to farming. It is very dry there. They only get about 10 to 12 inches of rainfall a year. I met with wheat farmers who wanted to improve the fertility of their land, but the problem was it is so dry that their yields are only about 20 bushels per acre. With wheat selling at $3/bushel, that does not leave much for a fertilizer budget. From their $60 per acre income they had to buy seed, plant the wheat, harvest the wheat, and make land payments. How could they do it? The answer is they don't grow wheat every year; they also raise sheep on those same fields. They plant wheat and let their sheep graze it the first year and second spring, and then in late summer they pull the animals out, let the wheat plants go to seed, and then harvest the wheat. It isn't possible for them to make any money just growing wheat, but by only planting once for both the sheep to graze and for wheat to harvest, they can make a little money off their land.

But what about their fertilizer budget? They have heavy soils that needed minerals. So I asked them, "If you only make $60/acre, what can you afford to spend on fertilizer?" "Well," they said, "we can give you about three dollars an acre." Were they going to bulk spread that three dollars worth of fertilizer? I told them if they were going to bulk spread it, we might as well go to the bar and have a beer and at least get something for the three dollars. Then we can go out to the field later and water something. We'll at least get a few plants started that way!

What the Western Australian wheat farmers did in the end was to put whatever they bought directly on the seed, focusing their fertilizer right on the germinating plant. They applied some natural nitrogen-producing organisms, a root stimulant and trace minerals, and saw a great response from a budget of just a couple of dollars an acre. The seed treatment doubled roots and cut water needs. The farmers were ecstatic over the response they saw.

Where To Put It

You should apply your fertilizer where it will do the most good. For the Western Australian farmers, putting their fertilizer directly on the seed was the best option — given their limited budget and harsh growing conditions. On my farm in Wisconsin, if I am going to make soil corrections, I bulk spread the fertilizer. I do the same for hay — it is pretty hard to put fertilizer in the row on alfalfa. But for my corn and other row crops I like to apply my starter fertilizer in a slot down the row, and I'll explain why.

When you put a mineral like potassium chloride in the soil, it goes instantly into solution. The positively charged potassium unhooks from the negatively charged chloride, and now you have potassium and chloride in the soil. Where do they go? There are four possibilities: the plants can take them up, they can hook to other compounds in the soil, microorganisms can consume them, or they can leach out of the soil. A lot of problems can occur when chloride hooks to available calcium in the soil and leaches. Another concern with applying highly soluble nutrients is that the sudden influx of ions like potassium and chloride can burn plant roots.

This is a good reason to buffer fertilizer. Of course, I don't use sources with a high salt index, but even with other fertilizers, buffering has its advantages. One way of buffering fertilizer is to place it in the soil away from the plant roots so there is soil between the roots and the fertilizer source. Another way to buffer fertilizer is to mix it with a carbon source. Rather than putting the fertilizer in the soil out of the way of the plant, mix something with the fertilizer so it has something to hook on to. That way less of the fertilizer will move into solution when it is applied, reducing the chance the fertilizer will burn the plant roots.

Buffer your fertilizer

How Fertilizer Burns Roots

We often talk about the potential for a fertilizer to "burn roots." High salt index fertilizers like 9-23-30 or potassium chloride don't actually burn roots like fire would burn paper. What they do is cause damage by pulling all the water out of the roots, dehydrating them, and potentially killing them.

When fertilizer goes into solution, it creates an area in the soil solution with lots of particles (called "solutes") in it; more than are in the root. Water will move across a membrane (the root wall) from an area of lower solute concentration to an area of higher solute concentration, which means water will move out of the root into the soil solution. This root dehydration is what is described as "burning roots." Burning roots means water gets pulled out of the roots because of the higher concentration of solutes outside of the roots, caused by the application of highly soluble fertilizer close to the roots in the soil.

Dr. William Albrecht, a soil science professor in the early 1900s and one of the founders of the idea of biological farming, said that the secret to fertilizer placement is to put the fertilizer into the soil so the roots can dodge the fertilizer. Put it into a slot four inches away from the row, or three inches over, or four inches down; bulk spread it on top; stick the fertilizer deeper down in the middle of the row; all so it cannot get near the plant roots. I like to place my fertilizer down below the roots in the middle of the row or else mixed in that zone. Exact placement does not really matter because as the plant pulls in water and minerals, the fertilizer will move in the direction of the roots. If you put your fertilizer into a band next to the plant, as the fertilizer moves into solution and starts to be taken up by the plant, it will migrate toward the plant root.

There are a lot of farmers, the zone tillage group, who believe very strongly in putting fertilizer in a zone in the soil. Instead of farming horizontally, they farm vertically. One farmer in northern Michigan, who farms this way went up against the university run field trials and beat the university by 40 bushels of corn an acre and 25 bushels of beans an acre. Zone tillage farmers mix some fertilizer in the root zone, and place other fertilizer, like nitrogen, deeper in the soil below the roots. They do this for two reasons: 1) so the fertilizer won't harm roots, and 2) so there is a higher concentration of minerals deeper in the soil where the roots can access them. Take

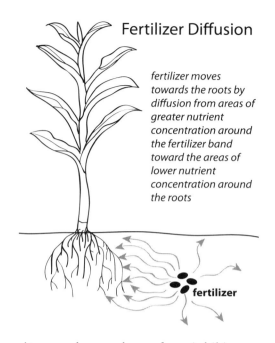

Fertilizer Diffusion

fertilizer moves towards the roots by diffusion from areas of greater nutrient concentration around the fertilizer band toward the areas of lower nutrient concentration around the roots

fertilizer

nitrogen as an example. Placing nitrogen close to the seed can inhibit root growth. If you put a lot of nitrogen in the soil where the roots can easily access it in the spring, roots will only grow deep enough to access that nitrogen. You end up with a tall, fast-growing plant with shallow roots. Later on in the growing season you are in trouble because your plants won't have big enough root systems to access water and nutrients when these are harder to find. By putting nitrogen deep in the soil, zone tillage farmers are encouraging their crop to develop a much bigger root system as the plant searches farther for more nitrogen. This leads to healthier plants that are better able to withstand drought and better able to access nutrients in the soil across the growing season.

Nutrient placement is not the only factor that contributes to root growth. Soil structure, soil moisture, and the level of soluble nutrients in the soil at planting time all influence root growth, but nutrient placement is one factor that can play an important role in root development.

When to Apply Fertilizers

On my farm I apply all of my soil correctives in the fall, and crop fertilizers at planting in the spring. An exception to this is my hay fields. If I am putting manure or potassium on hay ground, I put it on the last thing in the fall so it has a chance to work into the soil before I harvest the crop. I do not want a big flush of potassium in the spring because it will affect the health of my cows.

Corn Nutrient Uptake per Day by Growth Period

Nutrient uptake per day by growth period in corn				
Growth stage	Days after seeding	Pounds of uptake per acre per day		
		N	P	K
4 leaf	21	0.03	0.0006	0.04
9 leaf	34	1.14	0.1	1.4
Shoulder high	49	6.5	0.7	8.3
Tassel	71	3.8	0.7	4.0
Late silk	79	2.4	0.2	-1.1
Blister	93	0.2	0.4	-0.3
Dent	113	3.1	0.1	3.8

Source: Roots, Growth and Nutrient Uptake, Purdue University Department of Agronomy, Publication #AGRY-95-08

Applying a large dose of soluble nutrients in the spring does not fit with a plant's nutrient needs. The corn nutrient uptake table shows that for the first 34 days of a corn plant's life, it utilizes a total of only 18 pounds of nitrogen, 1.4 pounds of phosphorus, and 19 pounds of potassium. After 34 days the plant really takes off and starts growing, and the nutrient requirements of the plant then soar. At early tassel the corn plant has its peak nutrient needs, but this is close to 70 days after spring fertilizer was applied. If you applied only soluble fertilizers on your corn crop in the spring, very little of that fertilizer will still be available 70 days later (see also the chart "Phosphorus sources and uptake in corn," earlier in the chapter). Applying a blend of soluble and slow-release fertilizers in the spring helps to ensure nutrients will remain available across the entire

Small ant
Ammonium sulfate (Nitrogen)
Helps make other nutrients available.

growing season. Another option is to apply a smaller dose of nutrients in the spring, and go back out to the field and then apply a second dose later in the growing season when nutrient demands peak. But the real key to late-season nutrients is plant residue decay and soil biology. Having an ideal home for soil life ensures you will have healthy soil biology making nutrients plant available throughout the growing season.

What Fertilizer to Put On

In order to know which fertilizer to apply, you need to look at your soil test and the crop you are going to grow. The answers are individual, but an understanding of the principles of biological farming along with the advice of a good consultant will help you make these decisions.

How to Enhance Fertilizers

I like to do things to make my fertilizers work better. The first thing I do to enhance my fertilizer is add calcium. Calcium is the trucker of all minerals. Applying soluble calcium to the soil each year helps plants to access all nutrients, not just calcium. By applying more calcium, I have seen improved uptake of potassium, magnesium and trace minerals on my farm. I'm a strong believer in using calcium to boost crop production and plant health.

I also like to apply a little ammonium sulfate down the row to enhance the system. Ammonium sulfate acidifies the soil in the area around it, and this makes other nutrients more plant available. I try to limit purchased nitrogen because there are a lot of negatives to applying commercial nitrogen, but if you are farming biologically and have healthy, neutral pH soils, a little ammonium sulfate can help break down both naturally mined minerals in your fertilizer and less available nutrients in the soil. Since I farm organically I cannot use ammonium sulfate, but for a conventional farmer it is an excellent choice for enhancing the plant availability of other fertilizers, especially if they are from a less available naturally mined source.

Mixing nutrients with humates is another way to improve the efficiency of your fertilizer. Humates are a complex carbonaceous material that is very acidic. This acidity means that they work a little bit like ammonium sulfate to help break down naturally mined materials to make them more plant available. When you mix a mineral like high calcium lime or rock phosphate with humates and then apply the blend to the soil, the humates break down the rocks and you get a surge of minerals released into the soil. Humates have the added benefit of being a carbon source that can hold on to minerals, so not only will they help break down mined materials into a more plant-available form, they will also hook to the minerals in the fertilizer so they don't go into solution as quickly and tie up or leach.

When I first learned about humates, I got very excited and decided to try them on my farm. So I took 1,000 pounds per acre of humates and bulk spread it on top of the ground, then I planted my crop and applied my starter fertilizer down the row. I was looking for the surge in crop growth that had been described to me. I didn't see anything.

When I told the farmers who had recommended humates to me what happened on my farm, they laughed at me. They said, "If you've got 2 million pounds of soil on an acre of land, and you bulk spread 1,000 pounds of humates then apply fertilizer down the slot, how much of those humates came into contact with your fertilizer? Not very much! A speck or two. You need to mix the two together and **then** put them into the ground." Now that made a lot of sense. When I went back the next year and tried it again, this time mixing the humates with my fertilizer before application, I saw great results. I now use humates on my farm every year.

An Example From my Own Farm

The following is an example of what I typically do for a fertility program on my farm.

In the fall I put on soil corrective calcium. The amount I apply per acre varies based on my soil tests, but on a typical year I will bring in around nine semi-truckloads of calcium. If my farm was not certified organic I would use Bio-Cal, but since that is no longer an option, I use gypsum or OrganiCal. I plant my cover crops in the fall — often a blend of buckwheat, clover, oats, winter rye and hairy vetch. In the spring I work in the cover crops and then apply a starter fertilizer to all of my new crops. A typical blend would consist of the following:

5 pounds of zinc sulfate
10 pounds of magnesium sulfate
1 pound of actual copper from copper sulfate
1 pound of actual boron from sodium borate
100 pounds of HumaCal
50 pounds of Sul-Po-Mag
100 pounds of potassium sulfate
50 pounds of rock phosphate or chicken manure crumbles

One year a farmer came up to me after a presentation and said, "I had a consultant come to my farm and he made the same recommendations for all of my land. I know that's wrong. So I'm here today to ask, 'What do you do?'" My answer to him was that I use the same fertilizer for all of my crops. I have only one bin for all of my fertilizer. I do vary the amount of calcium and other soil correctives I put on per acre and I vary the amount of compost or manure I apply per acre, but the fertilizer is all the same — different amounts for different situations, but the same stuff.

The source of fertilizer you use on your farm has a huge impact on soil life and crop health. All nitrogen isn't the same. All potassium isn't the same. Just because the numbers on your fertilizer tag are bigger does not mean it is a better fertilizer. 9-23-20 is not a better deal than 10-9-10 just because there are bigger numbers on the tag. The source of the minerals in your fertilizer and what those minerals are hooked to plays a huge role in nutrient availability, nutrient tie up or leaching, and plant health. The best way to know if you are buying a quality fertilizer that will benefit your crop and not harm soil life or plant roots is to be informed. The more you know about the benefits and negatives of different fertilizers, the better off you will be when making a choice about the best fertilizer for your farm.

Calcium

Gypsum
or OrganiCal

Advancing Biological Farming

Chapter 9

Calcium

Over the years many people have accused me of overemphasizing calcium. I agree that I focus heavily on calcium, but I do not think it is an overemphasis — I believe that calcium is the key that makes everything else work. It plays a critical role in both plant and soil health. In the soil, calcium improves soil structure and water infiltration, and promotes abundant soil life. Calcium improves plant health by strengthening cell walls, adding stem flexibility, and helping plants cope with stress and respond to disease or insect attack. It also aids in the uptake of other minerals, which is why I call it "the trucker of all minerals."

I know I have already talked about the importance of calcium in previous chapters, but because it is such an essential component of a successful biological farming system, I felt it deserved its own chapter. Calcium really is the key element that improves every aspect of the biological farming system.

Calcium Functions

The following is a list of some of the functions calcium performs in the plant and in the soil:
- Improves root development
- Enhances microbial activity
- Increases the transport of minerals
- Improves soil structure
- Acts as a secondary messenger
- Improves plant health and disease resistance

- Is required for cell walls (pectin)
- Enhances the rate of protein synthesis
- Serves as a weed indicator

Calcium improves root development: Farmers cannot grow big healthy roots without calcium. A calcium-deficient plant will have brown, short, overly branched roots. If you take the boron and calcium out of your soil, the roots will stop growing immediately. Calcium stimulates the growth of roots and root hairs and is critical to the health and disease resistance of roots.

Calcium enhances microbial activity: Microbial activity in the soil is stimulated by calcium. It is an essential mineral for the growth and health of many types of soil organisms. Bacteria in particular thrive in high calcium soils because of the higher pH of soils that contain lots of calcium.

Calcium increases the transport of minerals: Calcium is the vehicle that moves minerals into plant cells. It is an integral part of the pumping mechanism of cell membranes that moves minerals from outside of the plant cell to inside of the plant cell. Without adequate calcium, mineral movement into plant cells would slow way down.

A number of years ago a consultant I work with compared feed test results from biological farms and conventional farms to see if the biological farms had more calcium in their feed. Most of the biological farms were applying Bio-Cal to their forage crops, so it was not a big surprise when he found close to 47 percent higher calcium levels in their alfalfa. What was interesting was that the levels of other minerals were higher too, even if the farmer did not add any of those minerals to the soil. It was the best proof I have seen that adding calcium to the soil improves uptake of all minerals, not just calcium.

Calcium improves soil structure: Soils that contain adequate calcium have a lot of aggregates (loose clumps of soil) and good tilth. Good tilth means the soil is well aerated, water infiltrates rather than running off, and the soil is able to retain water like a sponge. Good soil tilth also means roots can easily grow through the soil, seeds germinate well, and there is an abundance of beneficial microorganisms.

Research conducted by a USDA soil scientist confirms some of the beneficial effects calcium has on soil structure. The researcher looked at whether gypsum (calcium sulfate) could improve soil structure and water infiltration on no-till farms. To do this, the researcher set up three different treatments on small plots of soil. One of the soil plots had gypsum applied at the equivalent rate of four tons per acre, a second plot was bare soil with no fertilizer, and the third plot was no-till ground with residues on top. After the treatments were established, the researcher used a rain simulator to simulate rainfall of up to an inch an hour, and let it rain all day. All of the runoff water from each plot was collected so it could be examined to see how much soil was carried away and the size of the soil particles in the runoff.

On the no-till plot the runoff water looked dirty and there were a lot of fine particles in it. When the ground is hard and compacted on top, as no-till ground often is, rain takes off fine silt particles. By contrast, the water from the gypsum plot was clearer. The soil treated with gypsum was loose and crumbly, which allowed the rain water to infiltrate rather than run off. Of course, that was at an application rate of four tons an acre of gypsum, and most farmers can't afford to put on that much. If you apply just a little gypsum each year it will take longer to have this effect, but even small amounts of gypsum will help improve soil structure.

The other interesting thing the researcher found was that calcium from the gypsum did not just improve structure in the topsoil, it also infiltrated into the subsoil, which allowed crop roots to grow deeper and access more water. The plots with gypsum applied to them had better soil structure, better water-holding capacity, better water penetration and better soil tilth. This remarkable difference in soil structure was all due to the calcium and sulfur that were applied.

Calcium acts as a secondary messenger: Calcium acts as a messenger in plants, much like a hormone. When a plant is subjected to stress, like drought, heat, cold or a disease organism, calcium is released into the plant cells. This calcium signals the plant's defenses and helps it respond to stress.

A study conducted at the University of Wisconsin-Madison found that application of soluble calcium helped potato tubers survive heat stress. Potatoes are a cool-season crop and are very sensitive to heat. During the summer of 1988, the temperature was above 90 F for 46 days at the University's central Wisconsin research station. At the end of that growing season, potatoes in the plots where soluble calcium had been applied had

20 to 30 percent higher yields compared to plots where no calcium was added. The soluble calcium in the soil helped potatoes adapt to and survive heat stress.

One very dry year a farmer I work with called me out to his farm to look at something interesting on his cornfield. This farmer had done a lot of work with his soils and had beautiful soil tests, but on the majority of his cornfield the plants were stunted and brown and clearly suffering from water stress. However, there was one area on his field where the corn stood tall, green and healthy. The healthy corn made a funny looking circle on the edge of his cornfield, and outside of that circle the corn was stunted and dried up. The farmer said to me, "Do you know what that circle is? That's where I dumped my piles of lime and gypsum for the last several years." The farmer would load the lime or gypsum up off that area of the field to spread on the rest of his farm. This year he had planted corn where the gypsum and lime piles had been, and the extra calcium in the soil in that area made a dramatic difference in how the corn plants responded to the drought. If soil calcium levels were that high all over his farm, he would not have known there was a drought. Calcium made that big of a difference in how the corn responded to water stress.

Calcium improves plant health and disease resistance: In addition to helping plants cope with stress, calcium also helps plants stay healthy and resist disease. There are many published papers on how calcium works to protect legumes, fruit trees, vegetables, cereal crops and field crops from disease. These are plants of all different types, grown in a wide range of soil types and climate, and calcium plays a key role in disease resistance in all of them.

One way calcium helps plants resist disease is by strengthening both cell membranes and the plant cell wall. Many bacterial and fungal pathogens get inside a plant by secreting enzymes that break down the plant's cell wall. Once there is a hole in the cell wall, the pathogen can get into the plant and do damage. Calcium, however, strengthens the cell wall, so when the plant's calcium level is high enough it becomes very difficult for any disease organisms to breach the cell wall and get inside the plant. If a pathogen does attack, calcium acts as a messenger that alerts the plant to the attack so it can produce compounds to defend itself.

In Australia several years ago an article came out saying that our grandkids would not know what bananas taste like. Due to a fungal disease attacking banana plants, growers were worried that it would soon be impossible to grow bananas. The fungal disease called Yellow Sigatoka attacks banana leaves causing yellow and brown spots to form, eventually killing the leaves. Australian banana growers were applying fungicides every 14 days in an attempt to control Yellow Sigatoka. The fungicide applications were very expensive for the banana growers, to say nothing of the potential ill health effects for workers applying it and anyone who ate the fruit. After ten years of trying unsuccessfully to control this fungus, one grower decided to try a different approach. He applied calcium and boron as a way to make the plant stronger so it could resist the fungal attack. The difference was remarkable. The banana farmer saw total control of the fungus for 10 months, and was able to reduce fungicide applications to just four times a year.

After learning of the banana grower's success, an Australian researcher set out to look for a more widespread connection between calcium and boron levels in the soil and disease resistance. He started by testing soil at more than 50 sites in North Queensland, Australia, and then comparing disease rates. His study showed that there was less fungal disease on the higher calcium soils, and the least disease pressure where both calcium and boron levels were at adequate levels. His study, published in the *Australian Journal of Experimental Agriculture* in 2003, concluded that "Of the 10 sites which met the calcium criteria, nine sites had low levels of disease, while of the 40 sites which did not meet the calcium and/or boron criteria, 34 had high disease scores." Calcium levels in the soil, facilitated by boron, are directly related to the plant's ability to resist disease.

It has now been several years since the initial testing of calcium and boron as an alternative to fungicides on bananas. The growers continue to see excellent resistance to Yellow Sigatoka by regularly applying available calcium and boron to their banana plants.

The Important Role of Boron

When talking about the benefits of calcium, it is important to include boron in the discussion. Boron aids in the uptake of calcium. Based on years of trial and error and observation on my farm and other farms, I have seen the difference boron makes in getting calcium into plants. On my own farm, I never apply calcium without including boron in the blend.

Calcium is required for cell walls (pectins): Pectins, a part of the cell wall in plants, are long chains of sugar molecules that are linked together by calcium. They give cell walls their pliability, so plants with a lot of pectins have more flexible stems. They are also what make solid stems in alfalfa, and are highly digestible fiber for cattle. In addition, when pectin levels are high enough in the plant, it is a lot less likely that insects will attack the plant.

There is a direct correlation between calcium levels in the soil and pectin levels in the plant: If your soils are low in calcium, you will have less pectin in your plant. Green snap in corn, a condition where the corn stalk snaps off easily, is often blamed on genetics. That may be partially true, but elasticity in the plant has to do with how much pectin is in the cell wall. I apply calcium to all of my fields every year, and on my farm, regardless of the variety of corn I plant, I can take a corn plant and bend it over and touch the top to the ground. I also grow alfalfa with solid stems — that white spongy material that makes up the center of an alfalfa plant's stem is pectin. Having adequate plant-available calcium in the soil means your alfalfa stems will be stuffed full of calcium-containing pectin.

In the 1960s, Dr. Cary Reams came up with the idea that if you have a high enough sugar content in your plant, insects will not eat your crop. Dr. Reams believed there was a direct relationship between high sugar content (measured with a tool called a refractometer and reported as the Brix level) and reduced insect attack. He was convinced that if the Brix level from plant sap is over 12, meaning there is a lot of sugar in the plant, the bugs won't eat your crop because when an insect eats high sugar plant sap, the sugar turns into alcohol in the insect's body and kills it.

I am not sure if this is true or not. I have squeezed a lot of plant juice onto refractometers from a lot of different crops, and I do not always find a correlation between a high Brix reading and a lack of insect attack. I think that there is more to it than just the Brix reading. The refractometer doesn't measure just sugars; it measures soluble carbohydrates, which are a combination of sugars, starches and pectins. Starches and pectins are both comprised of long chains of sugar molecules, and I don't believe the refractometer can separate out one kind of sugar from another. If a Brix reading of 12 is due to a high level of pectins in your plant, then I agree the bugs won't eat your crop. But if it is due to starch or sugar, the bugs will still make a meal of it. I think the high pectin levels were protecting the plant from insect attack, not the high sugar content.

When Dr. Reams did his work linking Brix levels to insect attack, he was on the right track. He just needed to be able to separate out pectins from starch and sugar. It is pectins that protect the plant, not sugar or starch, though they may be indicators of a healthy plant that is actively photosynthesizing and producing a lot of sugars. It is difficult to measure pectins in plants, but it is possible to measure calcium levels. Since there is a correlation between calcium levels and pectin, if your plant has more calcium it will have more pectin, and therefore it will be more resistant to insect attack. I see this relationship between high levels of calcium, more plant pectin, and reduced insect attack in alfalfa, and I believe this relationship will be found in other crops as well, as studies into this continue.

As a plant gets more mature, the pectins convert into more complex carbohydrates: cellulose, hemicellulose and lignin. Those compounds make plants "woodier," which means they are tougher and harder for animals to digest. That is one reason why farmers like to cut hay earlier in the growing season — the plant will have a higher level of soluble carbohydrates, including pectins, and be easier for animals to digest. However, if your soil is ideally balanced and your calcium level is high, you can let the plants get older before you harvest them and you will still have high digestibility. That is why farms I work with that have well-balanced soils can let their hay grow longer before cutting it and still have good digestibility and good quality hay.

Calcium enhances the rate of protein synthesis: I believe there is a relationship between fertilizer and insect attack. While some soluble nitrogen is needed for healthy soils and crops, too much causes problems. There is a lot of research available that supports the idea that too much nitrogen in plants leads to insect attack.

When plants take up excess nitrogen from an overabundance of soluble nitrogen in the soil, free nitrogen levels go up in the plant and it produces more amino acids. Insects are attracted to crops that are high in free amino acids. In order to convert those free amino acids into complete proteins, the plant needs calcium and sulfur. If you have too much nitrogen and not enough sulfur and calcium, bugs will just devour your crop. But if those minerals are sufficient, the free amino acids can be converted into proteins, reducing the attractiveness of the plants to insects.

Calcium serves as a weed indicator: I have always said if you have a loose, crumbly, well-balanced soil you will have a hard time growing foxtail. Foxtail likes dead, hard, tight-packed soil that is low on calcium. I proved this to myself a few years back when I rented a farm that had such poor soils no one would rent it anymore. The farm had light sandy soil without a lot of minerals and a low pH. It desperately needed lime, but none of the previous renters would put lime on the land because they did not want to invest the money. When the owner approached me about renting the farm, I saw it as an opportunity to do some test plots on calcium products and soil building.

I took soil samples and then set up test strips of different calcium sources across the field with the lowest pH. When my crop started to come up, the field had a tremendous variation in it that was not related to my calcium test strips. There is a road that runs along the edge of that field that used to be gravel, but five years before I started working with the farm they blacktopped it. Close to the road the clover I planted came up, but farther from the road the field was pure foxtail with no sign of the clover I put in. So I went back and took another soil sample from right next to the road and one from farther away. Next to the road the pH was 7.0, farther from the road it was closer to 5.5. Why? My theory is that it was the gravel road. Gravel is limestone, and I think that the dust from the gravel kept the part of the field next to the road limed all those years. That's why I could grow a crop next to the road, but farther away from it, where the pH was low and the soil was lacking in calcium, the only thing that grew was foxtail. Foxtail does well on a hard, tight soil with a low pH and low calcium, but not much else does.

The Calcium Family

When I started working as a consultant, calcium was calcium, phosphorus was phosphorus, and nitrogen was nitrogen. I did not realize then that there are a lot of different sources of minerals, and they all do different things. One of the reasons mineral sources are all different is because no mineral is found by itself — you can't go out and buy pure calcium, it does not exist that way. Calcium is a positively charged element, and is always found hooked to a negatively charged element. There are hundreds of different compounds that calcium will hook to, which means there are hundreds of different ways calcium can be found in the world. Some of the calcium compounds make good liming agents, some make good fertilizers, while others are not useful for farming at all.

I already touched on some of these calcium sources in the chapter on fertilizers, but I wanted to discuss them again in more detail, and also go over research that has been done on different calcium products.

Calcium carbonate: Calcium carbonate is limestone, commonly known as high calcium lime. It can also be called calcitic limestone, quarry lime or ag lime. It generally runs about 35 to 39 percent calcium, and the rest of it is carbonate. I once ran into a salesman who told me he had a limestone product that was 98 percent calcium. I was embarrassed for him. There is no product out there that is 98 percent calcium. Calcium has to hook to something, and in limestone, calcium is hooked to carbonate. A ton of limestone has only two to five pounds of soluble calcium in it. It is a good source of calcium to use if your soils have a low pH. The carbonate in the limestone will help bring up the soil pH, and at the same time the calcium will become plant available as it is solubilized from the acidity of the soil.

Pel-lime is a type of calcium carbonate that has been very finely ground and pelleted. It is 35 percent calcium and very low in contaminants. The fineness of the grind makes the calcium more plant available. A fine grind gives the product more surface area for microorganisms to work on, allowing them to release available calcium into the soil. Pel-lime also has a binder in it, lignin sulfonate, to help keep it in pellets. Lignin sulfonate is derived from tree sap, and has the added benefit of being a biological stimulant that increases the availability of calcium and other minerals. Pel-lime works well on low pH, low calcium soils. It also works well if you can't get your calcium source worked into the ground.

Applying lime in the form of ground up rocks is a relatively new practice. When George Washington was a child they applied lime to soils, but there were no machines around that could crush stones small enough to use on farm fields. So where did they get their lime from? They burned it. If you go to visit some of the old farming museums in places like Maquoketa, Iowa, you can see the old lime kilns where they dropped big chunks of limestone down into fires and burned them. Why did they burn limestone? Limestone is calcium carbonate. "Carbonate" means there is carbon in it and any carbon source burns. Once the carbon is burned out of limestone, what is left is calcium oxide and calcium hydroxide, both of which are very soluble sources of calcium. That is what is in Bio-Cal. Bio-Cal is a soluble source of calcium, similar to the lime that was applied to the soil years ago. Ground limestone is a newer source of lime, and it is less plant available than burnt lime. Ground limestone works well on acid soils, but if you want additional calcium and the soil pH is near neutral or higher, burnt lime is a better choice.

Calcium magnesium carbonate: This is dolomitic limestone. Dolomitic limestone is generally a little browner in color than high calcium limestone. It's about 22 percent calcium and 8 to 15 percent magnesium, and has low solubility.

There is a lot of dolomitic limestone here in Wisconsin. Many farmers use it because it is an inexpensive, readily available lime source. Unfortunately much of the soil in Wisconsin is already high in magnesium, making dolomitic limestone a poor choice. The extra magnesium will tighten the soil, interfere with potassium uptake, and interfere with nitrogen sources. Also, if it is calcium I need, I get almost twice as much with high calcium lime as I do with dolomitic limestone.

Dolomitic limestone will bring up the pH of an acidic soil the same way high calcium limestone does. For this reason, dolomitic limestone is an affordable source of calcium and magnesium that works well on low pH soils that are also low in magnesium.

Calcium oxide: Calcium oxide is burned calcium carbonate, also called burnt lime or quicklime. It is 60 percent calcium, and with the carbon burned away, it is highly soluble and very reactive. When calcium oxide is mixed with water, it gives off heat and turns into calcium hydroxide. Because of this reaction, calcium oxide causes irritation if it is inhaled or comes into contact with skin or other moist tissues.

Calcium hydroxide: Water reacts with calcium oxide to form calcium hydroxide, also known as slaked lime or hydrated lime. It is 50 percent calcium, and like calcium oxide it can be very caustic. Calcium hydroxide is highly soluble and therefore a very plant-available source of calcium. You have to be very careful if you mix calcium oxide or calcium hydroxide with a nitrogen source like ammonium sulfate because it will react with ammonium and give off heat and ammonia gas. The smell can be overpowering.

If your soils have a low pH and are low in calcium, you apply limestone, but what if the soil needs calcium and the pH is above 6.8? Under those conditions, not a lot of the calcium from limestone will become plant available. If the soil pH is higher but you need calcium, calcium hydroxide can be a good source to use, especially if other compounds like sulfur are added to it to make it less caustic and friendlier to soil life.

Bio-Cal: A number of years ago I went in search of a good source of calcium for my farm. I have high magnesium and neutral pH soils, so the dolomitic limestone that is available in my area was not the right choice. After some searching, I found a small business in Iowa that had burnt lime from a cement kiln. It was very finely ground, micron fine, and it had a little sulfur, a little potassium and a few other things in it. I knew the fineness of the grind would make the calcium more available, so I figured I could justify my trucking expense by putting on a little less per acre. The sulfur and potassium in it offset some of the cost as well. What I did not realize was that this new product was not just finely ground high calcium lime, it was in fact burnt lime, and they are not the same thing. After going through the kiln, all the carbon had burned away and what I had was calcium oxide and calcium hydroxide. Burnt lime is fertilizer-grade calcium, not lime. When I applied it to my fields, I immediately saw a plant response, especially on my hay. My alfalfa plants were bigger, the leaves were larger, there were more pectins in the stem, and the tissue tests showed an increase in calcium the same year I applied this experimental product.

The experimental new product from Iowa eventually became Bio-Cal, and for years I put it on every acre of my farm and saw great results. Bio-Cal works wonders on legumes. Unfortunately, when I converted my farm to organic I could no longer use Bio-Cal. Because the mineral is burned, it is not approved for use on certified organic land because burning is a synthetic procedure and synthetics are not allowed in organics.

I did a research project a few years back with a conventional potato farmer using Bio-Cal on his potatoes. Potato growers struggle to get enough phosphorus into their plants. This farmer was spoon-feeding his crop phosphorus multiple times over the growing season and watching phosphorus levels in his potato crop go down, down, down. I started him on Bio-Cal, and we reduced the amount of phosphorus he was applying. The farmer watched his phosphorus levels go up and up in the plants early in the growing season, and then hold until the potatoes started to bulk up. The potato grower was really impressed. He said that for years he had been applying large quantities of soluble phosphorus, trying to get more phosphorus into the plant, but he was not getting anywhere. When we dropped the amount of phosphorus going on the crop and added Bio-Cal, we got more phosphorus into the plants. Calcium is the trucker of all minerals, so adding more phosphorus was not the solution to getting more phosphorus into his potatoes, adding calcium was. He had a huge reserve of phosphorus in his soil that the plants were not able to access, and in my opinion, the calcium we added stimulated soil biology, which in turn made some of that reserve phosphorus in the soil available to his potato plants.

Calcium sulfate: Calcium sulfate is commonly referred to as gypsum. It is 18 to 20 percent calcium and 17 percent sulfur; close to a 1:1 ratio of calcium to sulfur. There are 25 pounds of soluble calcium in a ton of gypsum, making it almost 100 times more soluble than limestone.

Gypsum works especially well as a soil corrective on soil that is high in magnesium or soil with an adequate pH that is tight and compacted. Gypsum is good for soil structure because it will break up soil crusting and open the soils for better water penetration. Farmers can put on up to four tons per acre each year, as long as the soil is high in magnesium. As already discussed, the sulfur in gypsum is an anion, and when it goes into solution in the soil, it combines with magnesium (a cation). Sulfur and magnesium form magnesium sulfate, Epsom salts, which are very plant available, but also leach easily from the soil. When soils are high in magnesium, applying gypsum will help reduce magnesium levels in the soil, but when magnesium levels in the soil are low, you do not want to overdo gypsum or there could be a magnesium deficiency. Gypsum is also not a good fit on low pH soils.

Spreading gypsum is a good way to reduce the ammonia smell on barn alleys. A lot of farmers put limestone on their alleys, but limestone actually releases ammonia and makes the smell worse in the barn. Gypsum reduces the smell because it ties up ammonia, rather than releasing it like limestone will.

Calcium phosphate: Calcium phosphate is rock phosphate. It is a source of both calcium and phosphorus, and is up to 30 percent phosphorus and 16 to 30 percent calcium. Because it is a mined mineral it has a lot of micronutrients in it as well, including rare earth minerals. Rare earth minerals are some of the 70 minerals needed to grow plants that are found in our soils in very, very minute quantities. They are removed from the soil in crops, and over time can become depleted. I believe that putting these rare earth minerals back in the soil will improve crop health and consequently crop yields.

Rock phosphate has a low solubility, which makes it a good source of both phosphorus and calcium for low pH soils. However, unlike limestone and dolomitic limestone, rock phosphate does not raise soil pH, so should not be used as a liming agent. Rock phosphate can be applied to neutral pH soils if it is worked into the upper layer of the soil where microorganisms can work on it to make it more plant available. For livestock farmers needing extra phosphorus, putting the rock phosphate in manures increases its solubility and provides more plant-available phosphorus.

Calcium nitrate and calcium chloride: I am not going to say much about these sources because I rarely use them or recommend them. They are both sources of very available, fertilizer-grade calcium, and they are often used as foliars because they are quite expensive. Potato growers often use calcium nitrate as a foliar spray because it is a good source of available calcium that also has nitrogen in it.

Composted chicken manure: Composted chicken manure from laying hens contains around eight percent calcium, five percent nitrogen, five percent phosphorus and three percent potassium. It is basically lime that has been run through a chicken. Kidding aside, commercial laying hens are fed a diet of around 10 percent lime, and much of that lime goes through the chicken and comes right back out again in the manure. The manure is then put into a pile and composted, which means in the end you get a good source of calcium that is tied to microorganisms and carbon.

Composted chicken manure is one of my preferred sources of fertilizer. It supplies calcium, nitrogen, phosphorus and potassium in a natural, plant-available form, which is not prone to leaching or runoff because it is composted. I like to apply 1,000 pounds per acre of composted chicken manure to my new seedings, Conservation Reserve Program (CRP) land, and to my rented acres.

Not all chicken manure from laying hens is composted. It is also possible to find sources that have been dried or are applied fresh. These are also good sources of calcium fertilizer, but the composted sources have the added benefit of being tied to stable carbon.

Calcium humates: When Bio-Cal could no longer be used on organic farms, I went in search of a replacement. I needed a source of available calcium that I could use on my organic farm, and that is why HumaCal was developed.

HumaCal is a blend of finely ground calcium and humates. Humates are a complex, carbonaceous material that are very acidic and have a huge capacity to hold minerals. When you mix humates with calcium, the humates adsorb the calcium, loosely holding it so it does not tie up with other compounds but can move easily into the plant-available pool of nutrients in the soil. HumaCal is a black, pelleted material that contains 18 percent calcium, 35 percent humic substances, and six percent sulfur. It provides available calcium for improved plant uptake and some sulfur in sulfate form. All of the ingredients in HumaCal are approved by the National Organic Standards Board, so it is okay to use on organic ground. HumaCal is a good calcium source to use on almost any soil type and because it is pelletized it will blend with other fertilizers for easier spreading.

The first load of pelleted HumaCal came to my farm for testing. Humates are notoriously dusty and difficult to handle, so by mixing them with calcium and sulfur and pelletizing them, I hoped not only to get the available calcium I wanted, but also to make the humates easier to handle. Unfortunately, that first load was pretty lousy. It was half dust and wasn't pelleted very well. When it was delivered to my farm, I told my son to spread 1,000 pounds per acre on the pasture. That is a large dose of HumaCal, and quite expensive to spread at that rate, but it was our first load and I wanted to test the new product to see how well it would perform. So I told him to put it on the pasture, but leave me some check strips so I could see how the new product would do. I got home that evening and there were no flags on the field. I asked my son where my check strips were. He said, "Well, if it's good enough for a little bit of the field, it's good enough for all of it." But what about my research? I like my test strips and my demonstration plots, but my son figured research plots were more work, and if the product was supposed to work well, why bother? "We'll know it worked on the pasture if the cattle eat it," he said. I didn't get any test strips in that first year, but over the years I have seen some great crop responses to HumaCal.

One year a potato farmer who works thousands of acres ordered several semi-truckloads of HumaCal, saying only that he wanted to apply it because he had been watching his neighbors. They applied calcium and he did not, and that year he was harvesting a lot of rotten potatoes and they were not. He knew calcium was part of the reason his neighbors' potato harvest was better than his, and he wanted to give it a try. The farmer won't say much about what he saw, but something must have worked because every year he orders more HumaCal.

With this interest in applying HumaCal on potatoes, I decided to do a study with Dr. Jiwan Palta of the University of Wisconsin-Madison, comparing HumaCal to gypsum as a calcium source for potatoes. In this study, calcium was applied at second hilling at a 100 lb./acre equivalent of either HumaCal or gypsum. When the potatoes were harvested in the fall, the plots treated with HumaCal had reduced incidence of hollow heart, internal brown spot, black spot bruise and brown center compared to the plots treated with gypsum. The total yield averaged over the three years of the study was very similar between the two treatments, but after the bad tubers were removed, yield on the HumaCal plots was higher than on the gypsum plots. The study concluded that applying HumaCal to potatoes reduced the amount of disease and the total number of bad tubers.

The Relationship Between Calcium and pH

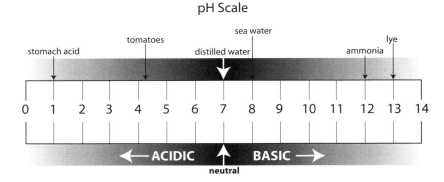

pH Scale

Acidity or alkalinity is measured on a pH scale from 0 to 14, with seven being neutral. Anything below seven is acidic, and anything above seven is basic.

Before starting a discussion about how to get more available calcium in the soil, I want to explain why lime is used on acidic soils to raise soil pH. The pH scale is a measure of hydrogen ions. When soil is acidic, it means there is an excess of hydrogen ions. In order to neutralize those hydrogen ions and raise the soil pH, the positively charged hydrogen ions need to hook to negatively charged substances in the soil.

When the pH gets below around 6.5, I recommend applying lime to raise the pH. Having a soil pH below 6.5 can have a detrimental effect on soil life, and also means that hydrogen ions are taking the place of positively charged nutrients on the clay particles and humus in the soil. This is why I say that when the pH is low, the soil is missing minerals.

Lime brings up the pH. When dolomitic limestone or high calcium lime is added to the soil, the positively charged calcium (or calcium and magnesium in the case of dolomitic limestone) will unhook from the negatively charged carbonate. The carbonate is now free to hook to positively charged hydrogen ions in the soil.

$$2H^+ + Ca^{2+} (CO_3)^{2-} \rightarrow Ca^{2+} + H_2O + CO_2$$

Hydrogen ion + calcium carbonate → calcium ion + water + carbon dioxide

As this equation shows, positively charged hydrogen ions will react with the negatively charged carbonate ion to form water and carbon dioxide. This takes some of the excess hydrogen out of the soil, thus reducing the level of acidity. In addition, calcium (and magnesium in the case of dolomitic limestone) is now unhooked from carbonate and free to either take the place of hydrogen ions on clay or humus in the soil, or get taken up by plants.

Since limestone is commonly used to raise soil pH, people assume that calcium raises pH, but it is not the calcium from limestone that raises the pH — it is the carbonate. It is also important to note that not all calcium compounds will raise pH. Rock phosphate (calcium phosphate) will have no effect on pH. The pH of an acidic soil goes up (becomes less acidic) only when the number of hydrogen ions in the soil is reduced, which happens when the positively charged hydrogen ions combine with negatively-charged compounds in the soil. When lime is applied, that negatively charged compound is carbonate.

Getting More Calcium into Plants

As discussed, calcium is needed for root growth, disease resistance, cell wall strength, drought resistance, soil tilth and many other factors in plant and soil health. I think it is pretty clear that without enough available calcium in the soil, crop health and yield will suffer. So how do you get enough plant-available calcium in the soil, and into your crop? Just having the right pH is no guarantee your crop is getting the calcium it needs.

Calcium is relatively immobile in the plant, meaning that once it is incorporated into plant tissues it stays there and does not move to other areas of the plant. This means that growing plants need a steady supply of calcium from the soil in order to meet their needs. Calcium already in the plant cannot be moved from the stem or the leaves to the seed during seed set. Calcium that is needed by the growing fruits or seeds has to come from the soil.

Feed Test Comparison between Conventional and Biological Farms

	Type of Agriculture		
	Conventional	Biological	% Difference
Crude Protein	19.39	20.5	6%
Calcium	1.10	1.62	47%
Phosphorus	0.31	0.36	16%
Magnesium	0.28	0.37	32%
Potassium	2.57	2.86	11%
Sulfur	0.24	0.31	29%

Based on feed tests from biological and conventional farms submitted to Dairyland Labs.

A consultant I work with, Mike Lovlien, compiled data from 19,000 feed samples that were sent to Dairyland Labs for analysis. Of those 19,000 samples, 295 came from biological farms, and the rest were from conventional farms from all over the Midwest. We wanted to compare calcium levels from the feed tests of biological farms that were applying calcium (most of them were using Bio-Cal) each year versus the feed tests of conventional farms that were not applying calcium. Our findings were even more remarkable than we had expected.

The data showed without doubt that a biological farming program results in higher levels of calcium in plants. The biological farms in this study had an average of 1.62 percent calcium on their feed tests compared to 1.10 percent on conventional farms. That was a difference of 47 percent more calcium in the alfalfa on the biological farms. But calcium was not the only mineral that was higher; phosphorus levels were 16 percent higher, magnesium levels were 32 percent higher, and sulfur levels were 29 percent higher. The biological farms were not applying potassium to their hay ground, yet the level of potassium in the alfalfa from biological farms was 11 percent higher than potassium in the feed of conventional farms. I am convinced that is because calcium improved biological activity in the soil and increased root growth, which resulted in more potassium in the plants.

Magnesium and potassium are both cations, and they are in competition in the soil. Normally, the more potassium you have in the soil, the more potassium gets into the plant and the less magnesium, but the biological farms in this study had more of both minerals in their alfalfa. How? They added calcium. In fact, all the minerals went up by adding calcium. That is why I call calcium the trucker of all minerals. Applying calcium gets more of all of the minerals into the plant, even if those minerals are not added to the soil.

The pH on these soils, both conventional and biological, was close to neutral, further evidence that having the correct soil pH does not necessarily guarantee plants are getting enough calcium. Otherwise, why were we able to get more calcium into plants by adding Bio-Cal?

This study clearly demonstrates that adding calcium to the soil changes plants. Of course, just because you put on calcium one year does not mean you will see this kind of response right away. The data from Dairyland Labs was from 295 farms we had worked with for a while, and these were average numbers from all of their feed tests. You may not see an immediate response, but applying calcium to your crop each year, from a source that fits your soils, will make a positive difference on your farm.

The farms in the above study were able to get more calcium into their alfalfa by applying Bio-Cal — but what about other calcium sources? After seeing the great response in plant calcium levels from applying Bio-Cal,

Total Amount of Cations in Solution after Application of Various Calcium Sources

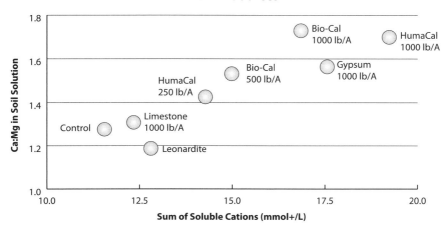

we were interested in learning how much available calcium is supplied by other calcium sources as well.

The Total Cations and Calcium chart shows the results of a study conducted at the University of Wisconsin-Madison with Dr. Phillip Barak. We added calcium from four different sources to a neutral pH soil, kept the soil damp and warm for 14 days, and then measured the soil solution to see how much calcium from the different sources was in a plant-available form. We also looked for an increase in other cations in solution to see if any of the calcium sources improved the availability of other nutrients. The four calcium sources used in the study were limestone, HumaCal, Bio-Cal and gypsum. Since HumaCal is a blend of calcium and humates (Leonardite), we also looked at straight Leonardite to try to see if the effect from HumaCal was from the humates or from the blend of humates and calcium.

The Total Cations and Calcium chart can be a little confusing. Basically, the higher up on the chart you go the more calcium is in solution, and the farther to the right you go, the more total cations are in solution. So the "control" dot at the far left has very little calcium and very few cations, while the "HumaCal at 1,000 lbs./acre" dot at the upper right has the most calcium and cations of all of the treatments.

It is interesting to compare the Leonardite (humates) treatment with HumaCal at 250 lbs./acre and HumaCal at 1,000 lbs./acre. The Leonardite treatment had very little available calcium, and not a lot of soluble cations, either. Compare that to HumaCal, which had a medium amount of soluble calcium and other cations at the 250 lbs./acre application rate, and the highest amount of available cations and second highest amount of available calcium at the 1,000 lbs./acre application rate. It seems that it is the combination of humates and calcium that makes HumaCal work to make calcium and other cations more plant available.

The Bio-Cal treatments had a lot of available calcium at both the 500 lbs./acre and 1,000 lbs./acre application rate, which was expected because Bio-Cal has highly soluble calcium hydroxide in it. It was not expected that gypsum would have so much available calcium. The 1,000 lbs./acre gypsum treatment had as much available calcium as 500 lbs./acre of Bio-Cal, and more soluble cations. However, the gypsum used for this study was not your typical quarried gypsum. It was 300 mesh, an extremely fine grind, and that fine grind makes the nutrients more soluble. I think if we had used quarried gypsum in this study the amount of available calcium and other cations from the gypsum would have been much lower.

It was very clear from this study that putting limestone on a neutral pH soil does not provide soluble calcium. A lot of people get confused and think that if you need calcium, you always add lime. If the soil pH is low, I agree, apply lime. But once your pH is neutral or higher, adding lime is not going to give you much plant-available calcium. On the chart, you can see that "limestone" is very close to "control." The control was just plain soil with nothing added to it. On the neutral pH soil used in this study, calcium from limestone was not any more available than adding no calcium at all.

The take home message from this study is that all calcium sources are not the same. If you have a neutral pH soil you need to find a calcium source with more available calcium than what is in limestone. Finely ground gypsum, Bio-Cal and HumaCal are all sources that provide soluble calcium.

A question that farmers often ask is, "My soil test shows I have plenty of calcium, so why should I apply more?" I always respond that just because your soil test shows you have enough calcium, that does not mean it is available to the plants. You need to have plant-available calcium to be sure your plants are not deficient. I apply calcium to my fields every year. I vary the amount and source based on my soil tests, but all of my fields get some calcium each year.

A study by Dr. Jiwan Palta, Professor of Horticulture at the University of Wisconsin-Madison, backs up the need for soluble calcium even when the soil test shows there is enough. Dr. Palta grew potatoes on a sandy soil that had 1,300 pounds per acre of calcium and compared that to potatoes grown on the same soil that received calcium nitrate fertilizer, a source of soluble calcium commonly used in the potato industry. Dr. Palta found that the potatoes grown on soil with adequate calcium but no additional added calcium suffered from heat stress, while the potatoes that received calcium nitrate did not. Applying calcium nitrate also improved disease resistance in the potatoes, reduced storage problems, and increased plant resistance to stress. It was very easy to see the difference between the potatoes that received calcium nitrate and those that did not. As calcium levels in the potato went up, the potato plant produced fewer, larger potatoes from healthier plants. Assuming that the plants were getting enough calcium from the soil did not achieve this result; adding soluble calcium did.

How Calcium is Lost From the Soil

What happens to available calcium in the soil? Some of it is taken up by plants and removed from the field in the form of crops. Some of it becomes fixed or unavailable in the soil. Some of it binds to other compounds in the soil and leaches. The formation of calcium chloride and calcium nitrate in the soil are the two biggest causes of calcium leaching.

Calcium chloride is formed when chloride from a source like potassium chloride fertilizer binds to available calcium in the soil to form calcium chloride. This removes calcium cations from the available calcium pool and ties them up with chloride, which means the calcium is no longer plant available. Calcium chloride can also leach, taking the calcium out of the soil.

A similar thing happens with nitrate from nitrogen fertilizers or manure. Calcium is lost from the plant-available pool when it binds to nitrate forming calcium nitrate. Excess nitrate from commercial fertilizer or dairy manure can wash a lot of the available calcium out the soil. I recommend that larger dairy farms always apply lime before spreading liquid manure. By applying a ton of lime per acre before applying manure, much of the calcium that will be lost as calcium nitrate is replaced. If you apply a lot of nitrogen from any source, you need to put on extra lime or you will end up with a reduction in the amount of available calcium in the soil. However, I am not sure how sustainable this practice is in the long run. Applying less soluble nitrogen each year is a better way to prevent the loss of available calcium from your soil.

Exchangeable Calcium and the Role of Boron

What all of these studies show is that what we are after is calcium that the plant can access — exchangeable calcium. In order to get exchangeable calcium you need a soluble calcium source. As mentioned before, limestone will work on low pH soils, but if your soil has a neutral pH, limestone is not a good option. Other sources, like gypsum, Bio-Cal, or finely ground calcium blended with humates all work much better.

Another very important aspect of getting enough calcium into plants is applying calcium with boron. I never apply one without the other. If calcium is the trucker of all minerals, then boron is the steering wheel. That is because without boron, calcium is not nearly as effective. Just drive down the road and ask every farmer you meet if they apply calcium without boron — very few of them will say yes. That is because farmers know from experience that their calcium will be a lot more effective if there is boron with it. That is why I add boron to all of my calcium sources.

Calcium is essential for healthy soils and healthy crops. It plays a central role in keeping soils loose and crumbly, improving air and water infiltration, and promoting soil life. Calcium provides benefits to every crop I know of by improving mineral uptake, strengthening cell walls, and protecting plants from stress and disease.

Conventional farmers who apply commercial nitrogen and do not add any calcium will struggle to get the quality and yield biological farmers get. Calcium, along with its partner boron, is such an important mineral for ensuring a healthy crop that I apply these minerals to my fields every year. Calcium is a keystone mineral for ensuring good soil structure and healthy crops.

So, yes, I place a lot of emphasis on calcium. I do not believe you can be a successful biological farmer without it!

Chapter 10

Mineral Interactions & the Big Four

Each of the minerals measured on your soil test plays a role in crop health and yield. Some of the minerals are essential for chlorophyll production and photosynthesis; some are essential for growth and protein production; some are needed for grain production; and some minerals are needed for moving other minerals around inside the plant to where they're needed. Not all of the minerals are needed in the same quantity, but they all have their roles.

When I apply fertilizers I include more than just nitrogen, potassium and phosphorus (NPK) in the blend. Crops remove a variety of different minerals from the soil, so it does not make sense to replace only nitrogen, potassium and phosphorus. Applying only the major minerals also is not sustainable. The soil has a certain ability to dish out nutrients, but eventually the minerals removed by crops will become limiting factors, yields will decline, and you will struggle to keep the soil and plants healthy. I don't want any minerals to be missing. Adding a balanced fertilizer provides the crop with nutrients above and beyond what the soil system can dish out. It prevents the soil nutrient levels from dropping lower and lower each year as crops remove nutrients that are not replenished. In crop production we have tremendous potential, and it isn't all about yield. Including a balance of minerals in my fertilizer helps to keep my crops and soils healthy.

The following graphic is a summary of the roles each mineral plays in a growing plant.

Benefits of Different Plant Nutrients

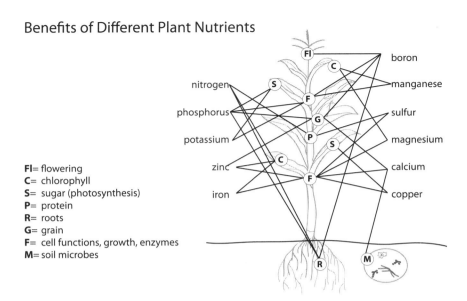

FI= flowering
C= chlorophyll
S= sugar (photosynthesis)
P= protein
R= roots
G= grain
F= cell functions, growth, enzymes
M= soil microbes

Nitrogen and sulfur are both part of the protein molecule. Nitrogen is also found in enzymes and the chlorophyll molecule.

Phosphorus is associated with energy because plants store energy in phosphate bonds. Phosphorus is also part of cell membranes and is important for healthy roots and crop quality.

Potassium is important for metabolism in the plant. It also plays a role in sugar movement within the plant, enzyme production, water balance and plant growth.

Zinc is an enzyme activator that is also important for growth and flowering.

Iron is required for production of the chlorophyll molecule, which is central to photosynthesis.

Boron is associated with root growth, cell elongation, plant health, and uptake of calcium.

Manganese is needed for photosynthesis and is an enzyme activator.

Magnesium is found at the center of the chlorophyll molecule, the plant's light-harvesting, energy-producing center. It also plays an important role in the production of oils and proteins, and in energy metabolism.

Calcium plays a central role in plant health and soil health. It has so many important functions, the previous chapter was devoted to talking about its many benefits.

Copper plays a role in cellular function and enzyme production, along with sugar production and sugar movement in the plant.

This is just a summary of the main functions of these minerals. They also have other vital roles in plants and in the soil, and these and other trace nutrients undoubtedly have other functions yet to be discovered.

Nutrient Interactions

There are many dynamic relationships between all of the minerals in the soil. When you apply a mineral to your field it does not just move into the soil solution and stay there in a plant-available form, waiting to get taken up by plant roots whenever the plant needs it. The soil is a very active environment and nothing stays in the soil solution for long. The minerals in the soil interact with each other, as well as with plant roots and soil life. The fertilizer you apply can make other nutrients more or less available, depending on how the nutrients added interact with other nutrients in the soil. The goal is to deliver adequate levels of a lot of different minerals to the plant, and understanding the relationships between these minerals helps you reach that goal.

The following is a list of some of the more common nutrient interactions that occur in the soil:

Nutrient Interactions

- Calcium and Magnesium

- Calcium and Boron

- Potassium and Sodium

- Potassium and Magnesium

- Zinc and Phosphorus

- Phosphorus and Aluminum, Iron and Manganese

- Phosphorus and Ammonium

Calcium and magnesium: Calcium and magnesium are both cations (positively charged elements). That means they do not hook to each other in the soil. Instead they compete with each other for sites on the clay/humus complex and for plant uptake. When your soil is high in magnesium it will be lower in calcium, and when the soil is high in calcium it will be lower in magnesium. When the soil gets too high in magnesium, it becomes difficult for plants to get adequate calcium.

The amount of magnesium and calcium in the soil also influences soil structure. Soil high in magnesium is harder and tighter than soil that is higher in calcium. Soils high in calcium tend to be loose and crumbly, and have good tilth — that chocolate cake-like texture that indicates a healthy soil.

I often run into trouble with researchers and conventional agronomists for saying that crops will be healthier if the soil has a high calcium to magnesium ratio. Many University researchers say that they have done a lot of work on calcium and magnesium levels in the soil, and the amount of calcium relative to the amount of magnesium does not make any difference. Any difference for what, I ask? I have seen their research and it does show that calcium to magnesium ratio makes no difference on yield when you put on optimum fertilizer. But what about quality? What about insect damage? What if I cut my fertilizer inputs in half? No one has ever explored that. When I ask why, they always reply that they already know how much fertilizer it takes to grow a crop, so why look at other factors? Most researchers do not look beyond the modern chemical-intensive farming system. They do know how much fertilizer it takes to grow a crop, but there are other factors involved in quality, yield and crop health that are rarely considered.

I think that what I say about having a good calcium to magnesium ratio in the soil often gets interpreted into "all you need to grow a healthy crop is a perfect soil test." But that is not what I am saying at all. My point is this, if the soil test comes back showing you have an excess of magnesium, why would you add more? And if we know soils higher in calcium relative to magnesium have better tilth, wouldn't you want to strive for that?

Recently a large conventional agriculture consulting organization came out with a recommendation that soil calcium to magnesium ratios should be five to one. That means you should have five times more calcium in your soil than magnesium. This company said that if you do not have that five to one ratio you are not going to have your best crops or your highest yields. I was very interested to see someone from the conventional world saying what I have been saying for years. I do not think you need to have exactly a five to one ratio of calcium to magnesium in the soil, but it is important to see the relationship between these two minerals.

Calcium and boron: Boron increases calcium uptake by plants. I include boron in all of my calcium sources, as do all of the farmers I work with.

A few years back I was going to do a research project with a University looking at whether we could decrease inputs and improve plant health by applying gypsum (calcium sulfate) and ammonium sulfate to potatoes. I wanted to add boron to the mineral blend in the study, but the researchers refused to do it. They said adding boron would add one more variable, and that they would not be able to analyze the research data if there were too many variables. In other words, was it the gypsum that caused healthier plants, or was it boron, or was it some combination of the two? I dropped out of the study after that. I did not see the point in going forward when I knew that applying gypsum without boron would not get the calcium uptake or the health benefits we were looking for.

It's very easy to see the difference in calcium uptake when boron is applied with a calcium source compared to when it is not. This has been proven by research, and I have seen it many times over the years by looking at tissue tests from farms that use boron and farms that don't. When calcium is applied without boron, there just is not as much calcium in the plant.

Potassium and sodium: Like magnesium and calcium, potassium and sodium are both cations. Since they are both positively charged, they do not hook to each other in the soil, and instead interact by competing with one another for plant uptake and for sites on the clay/humus complex in the soil.

Not too long ago I met a farmer with a cattle operation in Florida. Land down there is a lot more valuable than in the Midwest, so they farm a little differently. When land is worth $30,000 an acre, you either sell your farm and retire, or you have a lot of cattle per acre. In this case, they had 500 cows on 250 acres. They get 60 inches of rainfall a year and they irrigate when it is not raining. As you can imagine, keeping the pastures from turning into a muddy mess is a major problem. Another problem is what to do with all of the manure. If it is spread across the pastures, the cows won't graze — we all know that cows do not want to eat around the manure pile. They seem not to like the taste of the grass, and because it is high in potassium and nitrogen it causes health problems. On the Florida farm, the whole pasture is like the area around the manure pile. So how is the farmer going to get his cattle to graze?

On farms like this in some parts of the country it is standard practice to apply salt (sodium chloride) to the land. Since sodium and potassium are in competition in the soils, plants will take up less potassium when sodium is available. The cows graze better, and because the plants are lower in potassium the cattle also have fewer health problems. I have not tried it on my farm so I can't say firsthand whether or not it works, but I know a lot of farmers in the east who apply 100 lbs./acre of salt each year to their pastures to improve grazing.

Potassium and magnesium: These two minerals are also in competition, just like sodium and potassium, or magnesium and calcium. Potassium is one of the easiest minerals to get into the plant and magnesium is one of the hardest. If you apply a lot of soluble potassium, the plant will suck it up like it is drinking from a straw, and as that happens, less and less magnesium gets into the plant. This means that over-application of potassium can result in a magnesium deficiency. Adding more magnesium to the soil may seem like the obvious solution to this problem, but unfortunately most magnesium sources are not very plant available. Dolomitic limestone, for example, is an inexpensive source of magnesium but adding it won't fix a deficiency. The magnesium in dolomitic limestone is not very soluble, which means it takes a long time to become plant available. If I have a lot of potassium in my soil and I am having difficulty getting magnesium into the plant, the solution is to add sulfur. Sulfur will combine with magnesium to form magnesium sulfate, which is very plant available. The plant is then able to access more magnesium, which will address some of the problem caused by excess available potassium in the soil. It is also important to reduce potassium inputs so more magnesium can get into the plant.

One well-known problem that highlights the potassium/magnesium competition in the soil is grass tetany. Grass tetany, or the staggers in cattle, is caused by a magnesium deficiency. A sure way to end up with grass tetany in your herd is to turn them out onto pasture that is too high in potassium. The plants take up potassium at the expense of magnesium, so cows that eat the pasture will end up with a magnesium deficiency. Too much potassium, or not enough magnesium — they are essentially the same thing to the plant.

Zinc and phosphorus: Zinc is a cation (positively charged element) that interacts with negatively charged elements in the soil, like phosphorus. It is important to watch the amount of available phosphorus in the soil (P1 phosphorus on your soil test) relative to the amount of zinc because too much phosphorus inhibits zinc uptake. If P1 phosphorus levels on your soil test are more than 10 times the zinc levels, the plant won't be able to get enough zinc.

Phosphorus and aluminum, iron and manganese: Blueberry farmers grow their crop on low pH soils because there's a lot of iron, aluminum, and manganese available at a low pH. At a pH of 5.5 or lower, these three nutrients are plant available, and that's what the blueberry plants want. Now, you can grow blueberries on a seven pH soil as long as you have enough plant-available iron, aluminum and manganese. It is not the low pH that blueberries are after; it is the minerals that are available at that low pH.

Those same three minerals cause problems for phosphorus uptake on low pH soils. Iron, aluminum and manganese are strong cations and phosphorus is a strong anion. This means that if you apply soluble phosphorus to a low pH soil, the phosphorus will quickly tie up with iron, aluminum and manganese, and plants will have a difficult time getting enough phosphorus. Phosphorus accessibility can become a major limiting factor on acid soils. Adding lime to the soil can help to bring the pH up and help make phosphorus more available to plants growing on low pH soils. The problem with moving the pH up is that it limits the availability of iron, aluminum and manganese. If you have a crop that needs all of these minerals — iron, aluminum, manganese and phosphorus — the secret is to spoon feed the plant the minerals it cannot get enough of from the soil.

Phosphorus and ammonium: Rock phosphate is an excellent source of phosphorus, but on neutral or high pH soils, it can be difficult to make that phosphorus plant available. Applying ammonium from fertilizer or manure along with a phosphorus source helps make that phosphorus more available to plants. Acidity from the ammonium interacts with phosphorus to make it more soluble so plants can access it. On my farm, I like to apply manure after I have applied rock phosphate to help make the phosphorus soluble. Another way to make phosphorus more plant available is to add rock phosphate to your compost blend and apply that to the fields in the spring.

The Big Four

In the previous section, I discussed the interactions between minerals in the soil. This section is about getting four key minerals into plants. The soil level of the four minerals I call "The Big Four" isn't necessarily an indicator of how much of those minerals your plant can get. You need to take tissue tests or feed tests to see whether or not you have enough of these four minerals in the plant.

The idea of the Big Four first came about when I was struggling with our family's garden. We had a lot of insect and disease problems, and I started to wonder if there wasn't a relationship between the minerals in the soil and the health of the plants. Each spring when we took soil tests from our garden, the soil always seemed to be low on calcium, phosphorus and boron.

Another clue I had that something was going on between the soil and plants was in observations of dairy farms. Many of the farms I worked with noticed that their feed tests changed after they added minerals to the land. It was not very scientific, but I recall farmers saying, "Boy, this hay feeds well," after they had applied certain minerals to the land. So I knew something in those plants was changing, and that certain minerals made a bigger difference in feed quality than others.

After many years of looking at tissue tests and feed tests, I began to see a link between magnesium, phosphorus, calcium and boron, and plant quality. These minerals don't show up in plants by accident, and just buying them and adding them to the soil will not get them into plants. There is a biological link with all of these minerals, which is why I call them "indicators." The presence of these minerals in the plant indicates that something is going right in the soil. Tissue tests high in these four minerals always indicate very high-yielding, high-quality crops no matter what the crop is: corn, beans, alfalfa, wheat, rye, avocadoes, bananas or grapes. Whenever I see a tissue test showing above normal levels of all four of these minerals, I know that crop will be healthy and productive.

The Big Four

| Calcium | Phosphorus |
| Boron | Magnesium |

Calcium

There is no doubt that calcium plays a very important role in plant growth and health – that is why there is so much in this book about the benefits of calcium. Just to recap, some of the benefits include:

• When calcium levels in the plant go up, there is an increase in all other minerals as well, regardless of whether or not those minerals were added to the soil.

• Calcium is very important in producing cell membranes and plant pectins, which are an important part of cell walls. Having strong, healthy cell membranes and cell walls aids in a plant's ability to fight off insects and disease. Pectins are also a very important source of energy for cattle.

• Calcium acts as a plant messenger, allowing plants to respond to stressors like heat, cold and drought.

The ideal soil test model says that if you get 68 percent calcium and 12 percent magnesium on your soil test you will have adequate calcium in the soil for your crop. I have seen many farms that reached that perfect model of 68 percent calcium and 12 percent magnesium, but did not get enough calcium in the plants. Even though the soil test followed the model numbers, the crops were still low on calcium. There is also the philosophy that if the soil pH is 6.5 or above, that means there is adequate calcium in the soil. Time and again I have observed plants deficient in calcium growing on soils with a pH ranging from very low to very high. What I have learned from this is that there is more to getting calcium uptake into plants than what you see on your soil test.

Boron

Boron plays a role in plant growth, flowering and grain fill, yet its importance in crop health is often overlooked. The following story is a case in point. At a conference a man came up to me and told me he was a farmer from Western Australia, and I had been out to his farm when I was in that country two years earlier. He farms virgin land, which is flood irrigated. When I had visited his farm, he was trying to grow 250 bushels of corn per acre using Pioneer seed on 30-inch rows and applying up to 250 pounds of nitrogen an acre, but his yields were not that great. His corn ears were not filling out, and he had knobs on the corn ear ends, which he thought might indicate a nitrogen deficiency. When I went out to the field to look at his crop, the corn plants were green from the top right down to the bottom and even the ground was green. It was not a nitrogen problem — his corn crop had plenty of it. It was a boron problem. Seventy-five percent of the time, when you see knobs on the end of the corn ear it indicates a boron deficiency. The farmer said that after I left his farm, he went out and took tissue tests and confirmed that his crop was short of boron. He had been struggling for several years to grow 250 bushels of corn per acre, and once he added boron and a balanced fertilizer, he was able to drop his nitrogen inputs to less than 200 pounds per acre, and he reached his yield goal.

Calcium and Boron

Calcium and boron work together as a synergistic pair. I have often noticed that applying calcium by itself does not result in an increase in calcium levels in the plant. Factors like soil type and soil biology play an important role, but in my opinion the factor that makes the biggest difference in calcium uptake by plants is boron.

When I first started working as a consultant, I ran into a lot of farms with adequate soil test calcium yet they were struggling to get 1.1 percent calcium in their hay. For optimum plant and livestock health, I like to see calcium at around 1.5 percent. Applying limestone on these soils did not help, so I played around with different calcium sources to see if I could help the farmers get more calcium uptake. What I found was that applying a soluble calcium source (like Bio-Cal) with boron made a huge difference in calcium uptake. Calcium alone, even if it was a source of soluble calcium, just did not have the same effect. When I first observed this rapid increase in plant calcium levels when both boron and calcium were added to the soil, I kept a close eye on other minerals on those same crops to see if they would rise as well. Other minerals did go up, including magnesium, potassium, sulfur and phosphorus. This demonstrated to me that calcium and boron together really are key to nutrient uptake in plants.

I then started adding boron to all of my calcium fertilizers and I started seeing higher levels of all minerals in all kinds of crops. In alfalfa, as long as I did not overdo potassium, adding a calcium/boron mix produced solid-stemmed alfalfa that fed really well. If you look at conventionally fertilized alfalfa, you often see a big coarse, hollow stem that looks like a drinking straw. Compare that to a good biological farm where the alfalfa is fertilized with calcium and boron, and the alfalfa will have a solid stem, packed with pectins.

Based on years of observing plant response to calcium, I came up with the saying that calcium is the trucker of all minerals and boron is the steering wheel. Calcium drives the uptake of all minerals into the plant, and boron directs it. Applying calcium without boron is not nearly as effective as applying calcium *with* boron.

But a high calcium level in the plant was not the whole story because occasionally I would see a high calcium alfalfa crop that did not feed well. This got me thinking about other minerals as indicators of plant health. Calcium and boron are a big part of the picture, but they are not the whole picture.

Magnesium

Having sufficient magnesium is essential to plant growth because it is found at the center of the chlorophyll molecule — the molecule that captures the sun's energy. Plants without enough magnesium can't maximize photosynthesis, which means they do not grow as quickly or produce as well as they would with ample magnesium. Besides being necessary for plant growth, magnesium also plays an important role in the production of oils and proteins, and in energy metabolism. It is easy to spot a deficiency of this nutrient because plants start to get yellow streaks on their leaves from the lack of green-pigmented chlorophyll.

Magnesium is difficult to get into plants. I have seen many farms with high magnesium soils yet their crops are starved of magnesium. One reason this happens is because, as I have already mentioned, magnesium and potassium are on a teeter-totter in the soil, so uptake of one interferes with uptake of the other. This means that the more soluble potassium I apply to the soil, the less magnesium plants can get.

Applying sulfur to the soil is one way to get more plant-available magnesium. Sulfur will hook to magnesium in the soil forming magnesium sulfate (Epsom salts), which is very soluble and plant available. The other way to get more magnesium uptake into plants is through healthy soil biology. This is the main reason I include magnesium in my Big Four. High magnesium levels on a tissue test are an indicator of healthy soils. It shows me that the soil has not been over-tilled or compacted, has plenty of air and water for soil life to thrive, and has not been subjected to too many soluble nutrients or any harsh fertilizers.

Why Not the Big Five?

Since I am such a big promoter of including adequate sulfur in a fertilizer program, some of my colleagues wanted to know why there isn't a Big Five, with sulfur being the fifth mineral.

I do not think it's necessary to include sulfur with the Big Four because in order to get adequate magnesium uptake you almost have to use sulfur. In this way, magnesium in the plant serves as an indicator of sufficient sulfur as well as healthy soil biology.

Phosphorus

Phosphorus is important for photosynthesis, flowering, fruiting, nitrogen fixation and plant maturation. Having adequate levels of this nutrient enhances root growth, and it can also strengthen plant stems, reducing the risk of lodging. I want phosphorus to be active and cycling within my soil, not leaching, not eroding, and not getting tied up where plants cannot access it.

Phosphorus is probably the most difficult mineral to get into plants. It has a strong negative charge and ties up with other minerals in the soil very easily. It is difficult to maintain in the soil solution where it is available to plants, especially on high pH or low pH soils. Even on neutral pH soils, phosphorus availability goes down soon after it is applied. According to the University of Wisconsin, bulk spread DAP (diammonium phosphate) starts to tie up within two hours of spreading it. Because of this, applying soluble phosphorus to the soil does not guarantee you will get phosphorus into the plant. If your soils are compacted, low in organic matter and without much soil life, you are going to have difficulty getting phosphorus uptake. Without soil biology, and in particular mycorrhizae, it is difficult for plants to get enough phosphorus.

What are mycorrhizae? They are a type of filament-like fungus that forms a symbiotic relationship with plants. They colonize plant roots and grow outward to form a large network in the soil that captures water and nutrients and brings them to the plant. Mycorrhizae are best known for

How Mycorrhizae Increase Root Surface Area

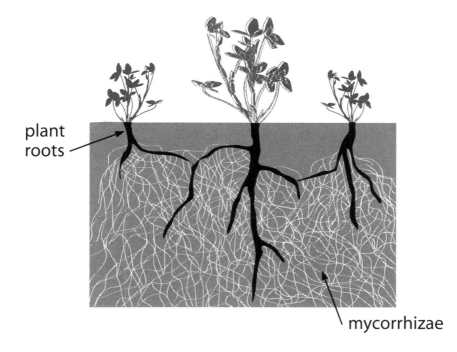

plant roots

mycorrhizae

bringing available phosphorus to plant roots, but they also capture zinc, manganese and copper. In exchange for delivering nutrients to the plant, the plant provides the mycorrhizae with carbohydrates and vitamins essential for their growth.

There is a feedback between the amount of soluble phosphorus in the soil and the amount of mycorrhizae that will grow. If a lot of soluble phosphorus is present in the soil, the plant won't feed the mycorrhizae because the plant will perceive that it has enough phosphorus and does not want to expend the extra energy. This creates a feedback loop: soluble phosphorus is added, and as a result fewer mycorrhizae are present, so even more soluble phosphorus is needed. A better way to get more phosphorus into plants is to encourage growth of mycorrhizal fungi.

On my own farm, I accomplish this by doing everything I can to promote soil biology. I keep my soluble nutrient levels low and grow cover crops to feed soil life and build organic matter. I avoid aggressively tilling and tearing up the middle zone where plant roots grow. Not only would tillage disturb soil structure, it would tear up any mycorrhizal networks that are growing there. By encouraging healthy soil life, if my crop needs more phosphorus, the plant will send more carbohydrates to its roots, feeding the mycorrhizae and encouraging them to grow. This in turn will induce the mycorrhizae to bring more phosphorus to the plant roots. Through encouraging the natural biological cycle on my own farm I have been able to increase plant tissue phosphorus levels without applying any extra phosphorus to the soil.

I know this system is working on my farm because I grow a lot of forages, and I have almost double the phosphorus uptake compared to state averages. This indicates that I have healthy, high-producing crops with lots of energy. I also now have the minerals in the feed, where I want them, and I have not had to supplement additional phosphorus to my cows in over two years.

If I see a tissue test phosphorus level that is really high, I know good things are happening. I know there is a lot of energy in the crop from all of the phosphorus in the plants, and if it is a forage crop I know it will feed well. I also know the soil is healthy and full of life, which is good not only for phosphorus uptake, but also for uptake of other nutrients and the health of the crop.

Plants need more than just nitrogen, phosphorus and potassium (NPK) to grow. In addition to the 16 minerals usually measured on a soil test, there are many other minerals that are needed by plants in very minute quantities. A shortage of any of these minerals can be a limiting factor on crop yield and health. That is why I recommend applying a balanced fertilizer including trace minerals, and using naturally mined minerals — to get the extra "goodies" — those less common minerals not found in most fertilizers. I want to feed my crop a balanced diet, and I do not want any minerals to be missing.

However, even if all the minerals a plant needs are there in the soil in adequate amounts, there is no guarantee the plant can access them. The soil is a complex, dynamic environment where minerals interact with each other, tie up, or are lost from the soil solution. Some nutrients are easy to get into a plant, and others are much more difficult. For example, if I go out and apply nitrogen to the soil, I can go back a week later and the plants will have dark green leaves. The plant sucks up nitrogen like it is coming through a straw, and I can literally see the results of applying it just days later. Magnesium is a different story. If the soil is high in potassium, or lacking in soil life, it is difficult for plants to access magnesium. Even if a magnesium source is applied, without a good balance of minerals and adequate soil biology, the plant will have a hard time getting that magnesium.

That is one reason I came up with the Big Four. These four minerals: calcium, boron, phosphorus and magnesium, are more difficult to get into plants. Over the years I have noticed that if I can get plant tissue levels high in all four of these minerals, I will have a healthy crop that is resistant to disease and insects. That is because The Big Four serve as barometers of healthy soil biology. Getting the Big Four minerals into a plant requires a plan and a program — they just do not show up without effort. That is one reason why I call them indicator minerals. They indicate the farmer did something to improve his soils, and that something is working.

Together, the Big Four let me know that I have healthy soils and a healthy crop. It does not matter what kind of crop it is, whenever I see a tissue test that is high in calcium, magnesium, phosphorus and boron, I know the crop that test came from will be one of the healthiest, highest-yielding crops around.

Chapter 11

The Role of Soil Life

Biological farming is farming with soil life. Biological farmers do everything they can to promote healthy soil biology — from avoiding harsh fertilizers, to growing cover crops, to practicing thoughtful tillage. Contrast that with conventional agriculture which is farming with chemistry. The amount of fertilizer needed is calculated based on the type of crop grown and the expected yield. Any weed, insect or fungus problems that come up are treated by dosing them with chemicals to get rid of the problem. Conventional agriculture does nothing to promote soil life because it is assumed that the soil life will take care of itself. It is similar to how conventional agriculture views calcium — if the pH is fine, the plant must be getting enough calcium. But over time, farming with chemicals alters soil biology. The soil life that survives in a conventional farming system will be there based on what was added to the soil, what food is available, and what can tolerate the farming practices. In many cases, that means a low diversity of organisms and fewer of the beneficial types of soil life.

The millions of creatures in the soil play an important role in making nutrients available to plants and in combating diseases and crop pests. Loose, crumbly soil full of life and organic matter provides available nutrients to your crop and improves both crop yield and health. Biological farming is farming for the long haul, and biology in the soil is the key to successful biological farming. Create an ideal home for soil life and provide lots of food of different varieties and wonderful things can happen.

In an earlier chapter, I talked about a farm I purchased that had really nice looking soil tests but grew poor crops. That farm is an excellent example of why there is more to farming than just chemistry. Based solely on nutrient levels, it should have been a high-yielding farm. Soil calcium levels were 67 to 70 percent of base saturation and magnesium was 20 to 25 percent of base saturation, pretty close to where I like to see them. Phosphorus and potassium levels were quite high, with available phosphorus (P1) at 100 ppm and potassium at 250 ppm, so I knew those nutrients were not going to be yield-limiting factors. Sulfur and micronutrient levels were on the low side, but each year I add those to my starter fertilizer and I knew I could make up for the deficiencies in the soil. All in all, the soil test suggested this would be a high-yielding farm. But that first spring when I sent a farm worker over to work up the soil with a rotovator, the land told a different story. The rotovator was hooked to a 175 horsepower tractor, and it did not have enough power to do the job. There were mudballs coming off the back, and the tractor struggled to run the rotovator. The ground was hard and tight and did not have much life in it. Just looking at the soil test did not tell the whole story. The soil test showed good chemistry, but it could not tell me anything about soil life. Biology plays a huge role in soil structure and soil health, and this farm did not have healthy soil biology. Even though the soil test looked very good, I did not get very good yields off that farm the first year.

The Benefits of Soil Life

The soil is full of life. There are many different types of creatures; from mites to earthworms; from fungi to bacteria, from amoeba to nematodes. And this soil life fills many roles; from eating debris, to preying on other organisms, to eating plant roots. But even with all of our modern scientific advances, scientists estimate that we have only classified about one percent of the organisms that live in that soil. That means that 99 percent of what is living in the soil has not been studied by science. We do not know anything about these organisms or what they do. What we do know is that each of the creatures in the soil has an important role to play. The organisms in the soil hold on to nutrients, bring nutrients to plants, suppress diseases, and keep plants healthy. They give soil its loose, crumbly, chocolate cake-like texture, and that warm, earthy, root-cellar smell (and that smell may be the best soil biology test available). Given how much we still have to learn

The Soil Food Web

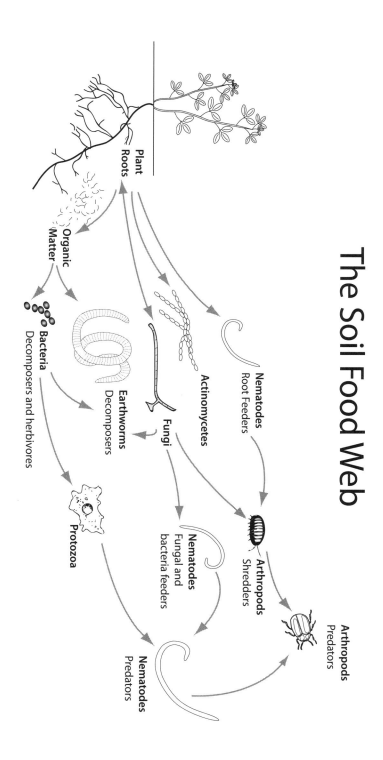

Plant Roots

Organic Matter

Bacteria
Decomposers and herbivores

Earthworms
Decomposers

Actinomycetes

Nematodes
Root Feeders

Fungi

Protozoa

Nematodes
Fungal and bacteria feeders

Arthropods
Shredders

Arthropods
Predators

Nematodes
Predators

about soil life, there are undoubtedly many other roles soil life plays that we have not yet discovered.

Soil life is important to farmers because it keeps the whole soil/plant system running smoothly. While we still have a lot to learn about what happens in the soil, we do know some of the benefits of healthy soil life, including:

- Suppressing disease
- Retaining nutrients
- Converting nutrients to a plant-available form
- Bringing nutrients and water to plant roots
- Opening channels in the soil for air and water infiltration
- Improving soil structure

Soil life suppresses disease: If the soil food web is working, no one population of organisms overpopulates and takes over. This includes disease organisms, which do not get a chance to ran amok and destroy a crop because there are other organisms in the soil that compete with them or eat them, keeping their population in check.

I saw a good example of competition between disease organisms and beneficial organisms when I was traveling in the tablelands of Australia, in north Queensland. The tablelands are the mountains above the rainforest, and they grow potatoes there. They are very close to the rainforest, which gets nine meters of rainfall a year (that's over 27 feet!). Sometimes they get two feet of rain in a single day. Boy, I know I would not want to farm in those flatlands above the rainforest. As you can imagine, with all that rainfall the humidity is extremely high. Molds and fungus are a big problem, and the farmers are forced to use a lot of fungicides. Most of them spray fungicides every five days to keep the molds in check. One year one of the farmers got tired of using so many fungicides, so he decided to try something different. He took a forty-acre hunk right in the middle of his farm and instead of applying fungicides, every five days he sprayed the potatoes with compost tea. Compost tea is a liquid full of microorganisms made by putting compost in a vat with water and brewing it until you have lots of microorganisms in the water (though it is difficult to say how many or what kind they are). The result was less mold and fungus on the 40-acre field where he sprayed compost tea than where he had sprayed fungicides.

Why? All microorganisms compete and when he sprayed on the compost tea the organisms from the compost tea established themselves on the leaves. In that situation, it meant that the problem microorganisms, the molds and fungus, did not get established.

The same thing happens in the soil. Organisms compete for space and food, and every organism has to defend itself against attackers. It is a dog eat dog world (or a bacteria eat bacteria world) in the soil. If you are a little creature in the soil, when something comes to attack you, you have to protect yourself. A common method of defense is to give off toxins to ward off predators. This is why antibiotics like streptomycin and penicillin come from the soil. Antibiotics kill bacteria, and when a soil bacteria attacks, some soil organisms give off antibiotics to defend themselves. Streptomycin and penicillin are just two of the compounds that we currently know something about. There are undoubtedly other antibiotics given off by soil organisms that we have not yet discovered.

Soil life retains nutrients: Soil life holds on to soil nutrients in a number of different ways. One of the most obvious is in the bodies of the microorganisms, earthworms, roots and other things living in the soil. Nutrients that are held in living organisms are not going to leach and be lost from the soil. They are temporarily tied up in living organisms, and when those organisms die and decompose, the nutrients again become available for other organisms to consume. Nutrients are also held in plant residues, though they are not immediately plant available because until they break down they are not in an available form. Soil life eats first, and once soil life has consumed those plant residues, they can release minerals and other plant foods back into the soil in a plant-available form. Highly decayed plant residues and soil life eventually turn into stable organic matter or humus. Humus contributes to nutrient retention because it has a high CEC, or cation exchange capacity, which is the nutrient-holding ability of the soil. More soil life and more humus means the soil has a higher CEC, and thus a greater ability to hold on to and dish out nutrients.

Soil life converts nutrients into a plant-available form: Only a small fraction of the nutrients in the soil are in a form immediately available to plants. A lot more of them are in forms that plants cannot access, and soil life plays a major role in converting them into a form plants can access. Nutrients that are made plant available by microorganisms include nitrogen, sulfur and manganese, among others.

Around 99 percent of the nitrogen in the soil is organic nitrogen, meaning the nitrogen is attached to carbon, either in the bodies of live soil organisms, in dead organisms, or in plant materials in various states of decay. Organic nitrogen has to be broken down and separated from carbon, or mineralized, in order for it to become plant available. This process is carried out by enzymes produced by microorganisms in the soil. The amount of nitrogen mineralized in an acre of soil each year varies widely depending upon the amount and type of organic matter in the soil, the climate, soil texture, and the health of the soil. Healthier soils have a greater capacity to mineralize nitrogen than lifeless soils, and warmer, wetter soils will mineralize more nitrogen than colder, drier soils. In addition, inputs like manure and green manure crop residues will mineralize rapidly, increasing the amount of nitrogen becoming plant available in the soil.

Growing a high-yielding, high-quality crop without the need for purchased nitrogen is a great measure of soil health. With healthy soil biology, you have now earned the right to purchase less nitrogen!

A similar system makes sulfur available to plants. Just like nitrogen, most of the sulfur in the soil is organic sulfur and needs to be mineralized by microorganisms in order to become plant available. And just like nitrogen, a healthy soil with lots of microorganisms in it will provide more plant-available sulfur than an unhealthy soil.

Rhizobia are a type of bacteria that are able to take N_2, the form of nitrogen in our atmosphere, and convert it to ammonium, which is a form of nitrogen that plants utilize. These bacteria colonize the root hairs of legumes forming root nodules. The plant and the bacteria form a symbiotic relationship where the host plant provides the bacteria with carbohydrates which the bacteria use for energy, and the bacteria supply the host plant with useable nitrogen. The amount of nitrogen produced in the root nodules varies based on climate and soil conditions, but it can be a very significant amount. Alfalfa, clover and vetch can all produce over 100 pounds of nitrogen per acre per year in good conditions.

Applying too much soluble nitrogen to the soil shuts down these natural nitrogen-fixing systems. You can easily test this yourself. Next year, go out and bulk spread 200 pounds per acre of ammonium nitrate on your soybeans, work it in the ground, plant your soybeans, and go back and check the roots a month later. How many nodules will you see? Not many. This is a very good example of how plants control soil life. Too much nitrate in the soil reduces the number of root nodules produced. It takes a lot of the plant's energy to supply the root nodules with carbohydrates, and if there is plentiful nitrate in the soil the plant won't invest the energy to supply its internal nitrogen-producing factory. Essentially, applying nitrogen makes the plants lazy, and once the plants get lazy, you have got trouble. The plants will use up the soluble nitrogen applied in the spring, and later in the growing season when that soluble nitrogen is not there anymore, the plant has not produced enough nodules to make adequate nitrogen for itself. It takes five pounds of nitrogen to produce a bushel of soybeans, and without sufficient root nodules, where will that nitrogen come from? Even though it sounds contradictory, by applying too much soluble nitrogen, your crop ends up short of nitrogen.

Earthworms are another source of plant-available nutrients. As they burrow into the soil, earthworms consume soil particles and organic matter, and excrete nutrient-rich casts. These casts are very rich in plant-available nutrients. One study found that compared to a silt loam soil, earthworm casts had three times more total nitrogen, four times more available calcium, three times more available potassium, and three times more available phosphorus. Interestingly, earthworms provide the most plant-available nutrients when they die. Their bodies are rich in plant-available nitrogen, phosphorus and sulfur, and when earthworm populations are large, as they die they contribute 50 to 90 pounds per acre of available nitrogen each year. In addition, the channels created by burrowing earthworms provide paths for roots to grow in a high-fertility environment.

Soil life brings nutrients and water to plant roots: Mycorrhizae are a type of fungus that forms a symbiotic relationship with plants. They colonize plant roots and grow out to form a huge net in the soil that captures water and nutrients and brings them to the plant. Mycorrhizae are best known for bringing available phosphorus to plant roots, but they also capture zinc, manganese and copper, and provide protection from pathogenic fungi and parasitic nematodes. In exchange, the plant provides the mycorrhizae with carbohydrates and vitamins essential for their growth.

In order to get a lot of phosphorus into plants, you must have mycorrhizae. Phosphorus ties up very easily with other minerals in the soil, which means that applying soluble phosphorus to the soil does not guarantee you will get high levels of this vital nutrient into the plant all season long. Biology, especially mycorrhizae, is the link between phosphorus in the soil and phosphorus in the plant. And just like applying too much soluble nitrogen results in fewer nitrogen-fixing root nodules, applying too much soluble phosphorus will result in plants not feeding the mycorrhizae, and not recovering as much phosphorus from the soil.

The proof of the pie is in the pudding: the proof that you have biologically active soils is high phosphorus levels in the plants, and you can see this for yourself on a tissue test.

Soil life opens channels in the soil for air and water movement: Earthworms are probably the best-known soil organisms on a farm. Because they are large enough to be seen by the naked eye, they are the easiest soil organisms to find and count. For many farmers, having abundant earthworms is a benchmark for healthy soils. One of their roles is to burrow into the soil, opening up channels that improve soil structure by allowing air and water to infiltrate. But earthworms are not the only burrowing organisms in the soil. Dung beetles, small mammals, and plant roots also provide channels that help aerate the soil and break up compaction.

Soil life improves soil structure: Soil life helps keep soil loose, crumbly and well-aerated with good water retention. I have already discussed how earthworms, dung beetles and other organisms open up channels into the soil for air and water movement. Another substance that helps improve soil structure is called "glomalin." It is a sticky material secreted by fungi that is very resistant to decay and plays a role in stabilizing soil aggregates, the loose clumps of soil that hold soil carbon and allow roots to grow. Having good aggregate stability gives soils that wonderful chocolate cake-like texture that is a sure sign of health. Glomalin is a key substance in the soil, but it was not even discovered until 1996. Who knows what other interesting and important things are under our feet, just waiting to be discovered?

Farming to Promote Soil Life

Soil life is a huge contributor to nutrient availability, good soil structure, and soil and plant health. As a biological farmer, you want soils that are rich with soil life. So what can you do to get more life in the soil on your farm? Farmers have to create an environment where the soil life can thrive, and farm to maximize the volume and diversity of soil life. We know soil life does well with the following conditions:

- A pH that isn't too low or too high
- Relatively stable soil temperature
- A carbon source to feed on
- Something growing on the land year-round
- A diversity of plants
- Thoughtful tillage

Soil life wants a pH that is not too low or too high: Different types of soil organisms prefer a higher or a lower pH. In general, bacteria prefer a slightly more acidic soil, which means a lower pH, while fungi prefer a slightly more basic soil, which means a higher pH. Earthworms do best at a neutral pH. If the pH is too far toward either extreme, none of the organisms that help your crops grow and access nutrients will do well. I like my soils to have a pH between 6.8 and 7.2 in order to have a balance of bacteria and fungi, and for earthworms to thrive.

Soil life wants a relatively stable soil temperature: Plants can survive a tremendous temperature variation aboveground, but not belowground. In order for the roots to survive, the temperature in the soil needs to stay relatively moderate. The same is true for soil life — it survives only in a very narrow temperature range. That is why I want to keep a blanket of growing plants or residues on top of my soil all year-round. Bare soil is hotter in summer and colder in winter. Keeping a blanket over the ground helps moderate the temperature so soil life and plant roots can thrive.

Soil life wants a carbon source to feed on: Soil organisms need food, and carbon is food for soil life. I like to feed soil life a blend of slow-digesting carbon from mature plants and the readily available, quick-digesting carbon from young green plants. The slower digesting materials like corn stalks and sawdust are fungal food while soft, green leafy materials from a green manure crop are bacterial food. Livestock manure or compost is another good source of carbon. And any time I can add additional carbon to my soil I do so because it is beneficial for soil life.

That is one reason why I like to keep my crop residues. After I harvest a crop like rye, wheat or corn, I need to decide what to do with the stalks. A lot of farmers I know bale the stalks and sell them, but I don't like to do that because then I am exporting carbon and nutrients off my farm. Leaving residues in the field allows that carbon to feed soil life and be returned to the soil. If I do remove the corn stalks or straw, I use it for bedding, compost it, and later return it to the soil. That way when my residues return to the soil they arrive predigested with goodies added.

Soil life wants something growing year-round: Take every opportunity to grow something green. Growing plants are a blanket for soil life as well as a food source. I do not like to leave land bare, because then there is nothing for the soil life to eat. My cows need to eat every day, and soil life is no different.

I like to plant a green manure crop in the fall, like winter rye, and work it into the soil in spring. The rye will start to grow in the fall and provide a blanket on the ground and food for soil life all winter long, and in the spring it will capture nutrients as it grows. I work it back into the soil when it is young and green, around one-foot tall, which means the soil life gets a big boost as it digests all that succulent green plant material.

Soil life wants a diversity of plants: All soil life needs food, and the greater the variety of food sources, the greater the variety of soil organisms. Many soil organisms live only on specific plants or types of plants, which means for each new plant I introduce, I am also introducing new types of soil life. As an example, look at *Rhizobia*, the bacteria that inoculate legume roots and fix atmospheric nitrogen. There are many types of *Rhizobia*, and each is specific to a small group of plants. The *Rhizobia* that live on soybeans grow only on soybeans and a few other plants, and the *Rhizobia* that live on clover grow only on clovers. There is not one organism that does it all. On the flip side, this means that every time I plant a new variety of legume, I am also feeding a different strain of nitrogen-fixing bacteria. The same is true of other types of plants and soil organisms — increasing diversity aboveground increases diversity belowground.

Soil life wants thoughtful tillage: It is important to conduct what I call "thoughtful tillage" to manage soil air, water and residues. The way to avoid soil compaction and get air and water movement in the soil is to till. Shallow incorporating of residues feeds soil life, and helps to keep the soil loose and crumbly so air and water can infiltrate and soil life can breathe. But do not till aggressively unless you have to. There are only two times when I think aggressive tillage is necessary: if you have a hardpan that air, water and roots cannot penetrate; or when you need to mix nutrients into the soil, such as when making a major soil correction or applying livestock manure. Other than those times, try not to disturb the root zone, where roots and soil life build complex networks that are easily destroyed by tillage.

Another reason not to till in the root zone is to protect the rhizosphere; the area in the soil right around the plant roots. The highest level of biological activity occurs here because this is where microorganisms gather to consume the carbohydrates and vitamins given off by the plant roots. Even in a drought, the rhizosphere will be damp because of all of the interactions occurring in this zone between roots and soil life. Protecting the rhizosphere is one reason why I shallow incorporate residues rather than work them deeper into the soil. I want to leave the mid-level root zone intact so I do not disturb the fertile, biologically active zone around the root channels.

An Example of Promoting Soil Life

I have seen a lot of farms produce beautiful, healthy crops by farming to promote soil life. An interesting example comes from a monastery in Iowa. Several of the monks attended my field day one year, and one of them came up to me and said, "In the Bible it says to leave land to lay fallow every seven years, and after visiting your field day we realized that we're the ones following the Bible, but you're the one leaving land to lay fallow and building soils." When the monks returned to their monastery, they laid out a seven-year plan for their farm that included one year of soil building.

The monks start out their fallow year by planting three bushels of oats per acre. When the oats get knee high, they turn them into the soil and plant three bushels of corn per acre. Rather than spend a lot on seed, they bulk spread the grain from their stored cob corn. When the corn gets five feet high they work it into the soil.

Dealing with Green Manure Crop Residues

Later one of the Iowa monastery monks wrote with an interesting question, "What should we do with all the residues from a five-foot tall green manure corn crop?"

I replied, you can't leave that much plant material lying on top of the soil because it will take too long to break down and a lot of the carbon will be oxidized and lost to the atmosphere. It's also difficult to incorporate that much residue just a few inches deep because with that much material, you end up just burying it.

The more residues there are, the deeper you can till to mix them in. That way the residues get dispersed deeper into the aerobic zone where microbes can work on them and break them down in order to release nutrients and build organic matter.

In this case, I recommended that the corn plants be worked into the soil about six to eight inches deep so they could break down quickly. Tilling that deep on bare soil would really be hard on soil life, but working up the soil when there are a lot of residues won't hurt the soil.

After the corn started to break down the monks planted three bushels per acre of soybeans from the bin and left it on the fields over the winter. In the spring they worked in the soybean plant residues and planted their crop. It was a bumper crop and weed free. The monks said they had the best crops with the fewest weeds the three years following the fallow year. After that, tillage and working the land start to degrade the soil again.

How Do You Know If You Have Soil Life?

There are tests available that measure microbiological activity in the soil. I ran some of these tests on my farm a few years ago, but when the results came back I really did not know what to make of them. If the numbers of bacteria or fungi on the test are low, is that bad? If so, what do I do about it? I did not feel that the test helped me make any management decisions on my farm. I think keen observation of the soil works just as well.

Soil life is something you can see, smell and touch. Dig down in the ground, rub some soil between your fingers, and feel the soil texture. Is it blocky or platey, sticky or gritty? You can see if the ground is crusted over and feel if the soil is tight and sticky. If you have ground that is sticky and too hard, it could be a problem with too much magnesium, but it is likely that you do not have enough air down in the soil so roots can grow down deep and keep the soil loose. Is it difficult to get a shovel into the soil, or can you easily push it in? Tight soil could indicate a problem with compaction. Dig out a few shovelsful and count the earthworms. The best time to do this on my farm is either at the end of May or in early September because earthworms are more active in spring and fall. (The best time to do this on your farm will vary depending on the climate where you live.) I like to see about 25 earthworms per square foot, digging channels into the soil and keeping everything aerated. Also smell the soil: does it smell earthy, or does it smell putrid, like a rotten egg? Your nose can tell you a lot about whether you have waterlogged soils with poor air and water movement, or healthy, aerated soils. Lots of earthworms, loose, crumbly soil, and that root-cellar earthy smell are all indicators of good biological function.

How I Promote Soil Life on my Farm

On my own farm I never plant a straight corn/bean rotation. After a couple years of corn and soybeans, I plant rye to give the field a rest. I put on 1,000 pounds of pelleted chicken manure for extra nitrogen and plant food for the rye crop, because not only do I want a decent rye harvest, I also want lots of biomass and carbon to feed soil life. When I harvest the rye, I usually leave the straw on the field because it has a lot of value as a complex carbon for feeding soil life. If I do take any stalks off for bedding, I put compost or manure back on to replace the biomass I have taken off.

I often interseed clover into my rye. After the rye is harvested in early August, the clover gets exposed to full sunlight and grows rapidly. By the middle of August, if there has been enough rain, the clover will be a foot tall and I will go out and rotovate it down.

Following that, I add soil correctives and subsoil if needed. Depending on the soil tests, I add either 1,000 or 2,000 pounds per acre of lime, gypsum or OrganiCal (a very finely ground blend of high calcium lime, sulfur, boron and humates). If the ground is tight, I subsoil; it is essential to manage air and water, and I do not want to have any problems with compaction. After my fall soil correctives are added and I have finished any subsoiling, I drill in buckwheat, which is one of my favorite green manure crops. It has an acidic root, draws a lot of phosphorus out of the soil, and produces a lot of biomass. If I have leftover corn seed, I may plant some of it as a cover crop as well. I let the buckwheat grow until it freezes. Depending on my plans for the field in the spring, after a hard frost I might leave the buckwheat on the field, or I might turn it back into the ground and plant winter rye. If I am planting corn or soybeans the following spring, I want the rye on as my winter cover crop, as it will come back in the spring. I turn it down when it gets a foot tall if I am planting corn, and I let it get a little taller before working it into the soil if I am planting soybeans.

By following this soil-building program, I am investing extra time and money in the field up front, but I make up for it in the long run because the crops that follow have a much greater potential. I have also just put carbon back in soil. I have aerated the ground. I have added nutrients and soil correctives. I have fed the soil biology by planting a diversity of different plants and working them back into the soil. I put protection on the ground over the winter, and next spring I get one more pass to work a young green manure crop in to provide food for the soil biology and nutrients for my growing crop. I am confident that my next crop will be very healthy and won't need as many inputs. If you are a conventional farmer, when following a soil building year you should not need any insecticide and you should be able to cut your nitrogen application in half. All farmers will find that weed control gets easier after soil building. Besides better yields, in the long run you will make a profit simply by reducing costs in other areas. As the soil becomes healthier, more biologically active, and better aerated, you will not need as many inputs to get what you have been getting, you will not have as many weeds, and you will have a better quality crop which generally yields better as well.

Living organisms in the soil have many roles. They decompose crop residue, manure and dead organisms in the soil. They store nutrients in their bodies, preventing those nutrients from leaching. Larger organisms will prey on crop pests, while smaller ones will prey on pathogens, helping to keep plants healthy. Some fungi form nets in the soil that improve soil aggregation or nutrient uptake, while earthworms and other burrowing organisms open up channels for air and water infiltration. Still other organisms change the form of minerals, making them plant available, or work together with plants to provide them with nutrients. A healthy, biologically active soil has millions and millions of organisms in it, all playing a role in providing nutrients to plants and keeping plants healthy.

As a biological farmer, I want to tap into the many benefits soil life provides. I want to do everything I can to create an environment where soil life will thrive. On my farm, this includes feeding the soil life a diversity of plants, and never leaving the soil bare. I take every opportunity I can to grow cover crops or increase diversity by planting a variety of crops and cover crops. I work residues back into the soil and apply manure and compost in order to provide carbon to soil organisms. I avoid harsh chemicals that could harm soil life, and I avoid tilling in the root zone as much as possible so I do not disturb fungal networks and plant roots. I also take stock of how soil organisms are doing by digging into the soil, feeling it, smelling it, and looking for signs of healthy soil life. If it looks like the soil is not healthy, I give the land a break and spend time soil building in order to rejuvenate soil life. Healthy soil life is a cornerstone of biological farming, and as a biological farmer, I do everything I can to nurture soil life on my farm.

Chapter 12

Soil Carbon

Most of the carbon in the soil is tied to living and dead organisms, so it is referred to as organic carbon, or soil organic matter. There are other forms of carbon in the soil not associated with soil life, like the carbon found in limestone (calcium carbonate). Since it is not associated with living organisms it is called inorganic carbon. This chapter is only about the carbon associated with living organisms: the soil organic matter.

Despite its importance, organic matter makes up only about 1 to 5 percent of the soil. It is made up of all of the living soil organisms, all of the dead soil organisms, and everything that was once living in the soil (plant and animal) and is now highly decomposed. I sometimes call it the living, the dead, and the very dead.

The living organisms present in our soils are important for improving soil structure, retaining soil nutrients, and converting nutrients to a plant-available form. I focused on living organisms in the soil in the previous chapter. In this chapter, I'm going to discuss the recently dead, or active part of organic matter, and the very dead, which is stable organic matter, or humus.

The active fraction of organic matter is where the nutrients are. It is made up of fresh plant residues, recently dead roots, microorganisms, earthworms and other recently dead soil life, and manure. All of these materials serve as food for living soil organisms. They also provide nutrients for growing plants, because as this recently dead material is consumed and broken down, plant-available nutrients are released in the soil.

Components of Organic Matter

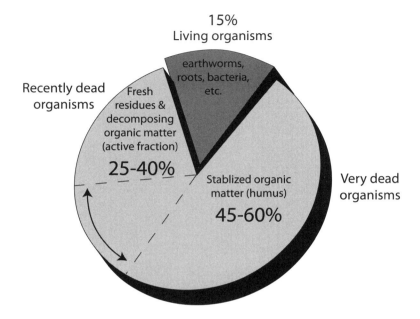

15%
Living organisms

earthworms, roots, bacteria, etc.

Recently dead organisms

Fresh residues & decomposing organic matter (active fraction)

25-40%

Stablized organic matter (humus)

45-60%

Very dead organisms

Highly decomposed materials in the soil become humus. Humus is formed as different microbes consume organic matter, and through many stages of decomposition, convert once living materials into very stable carbon, or humus. Compost is a good example of an intermediate stage in this process: compost consists of partially decomposed organic matter and humus. As the process of converting organic matter to humus proceeds, different intermediate chemicals are present at different points in time. Complete conversion to humus will eventually take place if organic matter is allowed to decompose for long enough under the right conditions.

Humus is highly decomposed and low in nutrients, and therefore is not a food source for soil life or plants. Even so, it plays a very important role in maintaining healthy plants and soil life. Humus has a very high CEC, or cation exchange capacity, which means it has a lot of sites on it that loosely hold nutrients, storing them for slow release to plants. Humus can chelate, or hold on to, harmful materials in the soil, neutralizing them so that they cannot cause damage to soil life or plants. It also improves soil structure, reducing compaction and improving the soil's water-holding capacity.

The Importance of Organic Matter in Nutrient Exchange

Humus greatly increases the mineral-holding capacity of soil. I often talk about minerals in the soil being held on the clay/humus complex. That is because both clay and humus have a lot of negatively charged sites on their surfaces that will hold on to positively charged minerals, or cations. That is where the term cation exchange capacity (CEC) comes from: substances that hold a lot of cations have a higher CEC than substances that hold fewer cations. Sandy soil, for example, has a low CEC, which means it has a low capacity to hold minerals. Clay is at the other end of the spectrum, which means that the more clay there is in the soil, the higher the CEC and the more capacity that soil has to hold minerals. Humus has an even higher CEC than clay. A sandy soil will typically have a CEC between 5 and 10, a clay soil can have a CEC of between 20 and 50, and humus has a CEC of between 100 and 300. That high CEC means humus has a tremendous capacity to hold minerals. Soils high in humus are very fertile soils.

Benefits of Organic Matter

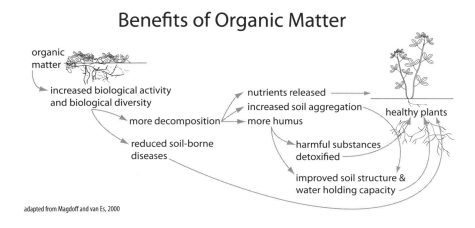

organic matter

- increased biological activity and biological diversity
  - more decomposition
- reduced soil-borne diseases

nutrients released
increased soil aggregation
more humus

harmful substances detoxified

improved soil structure & water holding capacity

healthy plants

adapted from Magdoff and van Es, 2000

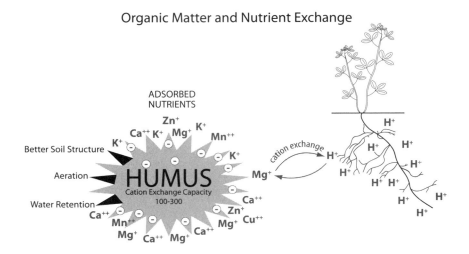

Organic Matter and Nutrient Exchange

Nutrients held on humus are in a form that is easily exchangeable with plant roots. Unlike nutrients that are tied up with other compounds in the soil, or part of the structural components of soil, nutrients associated with humus are held very loosely. Nutrients held by humus are "adsorbed," which means the nutrients don't form a tight bond with the humus; they adhere weakly to the humus molecules allowing the nutrients to move easily into the soil solution where they can be taken up by plants. Increasing soil humus by itself doesn't necessarily mean you'll have more available nutrients. The nutrients have to be present as well. If the soil is well mineralized, having a lot of humus ensures that many of those nutrients are held in a form that plants can easily access.

What Depletes Soil Organic Matter?

We call humus "stable organic matter" because it stays in the soil for a very long time. But even though humus is very stable, it can be lost from the soil. Under normal soil conditions, about two to five percent of soil humus breaks down each year. In a natural ecosystem or a balanced farming system, that amount of humus or more is replaced each year by the natural decay processes in the soil. The soil stays in balance and does not suffer a net loss of humus.

When chemical nitrogen and strong bases such as anhydrous ammonia are applied to the soil, they break down humus and it is lost. As the humus breaks down, carbon and the nutrients held on the humus move rapidly into solution. As this occurs, plants can get a big boost in growth from the sudden increase in soluble nutrients, but this benefit is short lived. The humus that is solubilized is lost forever, and building new humus in the soil takes a long time and good soil management. Farmers who use large amounts of chemical nitrogen and anhydrous ammonia are gradually losing humus from their soils.

Another way organic matter is lost from the soil is through too much tillage. Aggressive tillage can lead to loss of topsoil by wind and water erosion, and it also speeds the breakdown of organic matter in the soil. That is because turning over the soil through tillage exposes soil life to the air. Most soil organisms require oxygen, and as the soil is turned over, microorganisms are exposed to oxygen, resulting in a boost in microbiological activity. These microorganisms consume carbon from organic matter as their food source, so as their activity increases, so does the break down of soil organic matter. As they eat and digest, these soil organisms respire, which means they are giving off carbon dioxide as a waste product — releasing carbon from the soil into the air. This leads to a reduction in the amount of total carbon in the soil, and also a reduction in the amount of humus as less organic matter is present to be converted to humus.

Over time, loss of organic matter leads to compaction, lower soil water-holding capacity, poor soil health, and ultimately lowered crop health and yields.

Building Organic Matter in the Soil

Plants are composed of 40 to 45 percent carbon. As plants lose leaves and roots and eventually die, some of their carbon and nutrients (along with that from the soil organisms that consume the plants) becomes available to the next crop. Some of the plant tissue carbon is lost to the atmosphere as carbon dioxide when soil organisms consume it and respire, and some of the carbon stays in the soil where it can become stable organic matter. Most people assume that the way to increase soil organic matter is by

Characteristics of Carbon Inputs

	CHARACTERISTICS	CARBON TO NITROGEN RATIO	SOURCES
Green Carbon	Found in young living plants Builds very little humus Feeds primarily bacteria Rapid nutrient release	Low C:N ratio High N, Low lignin	Pig and chicken manures Young cover crops
Brown Carbon	Found in older, woodier plants Builds humus Feeds primarily soil fungi Slower nutrient availability	High C:N ratio Low N, High lignin Breaks down more quickly with added N	Lignified or brown plant material: corn stalks, mature cover crops Manure with bedding Manure from ruminants fed roughage
Black Carbon	Rich black material in soils Forms slowly over time Can be lost with excess tillage or too much applied N	High in complex carbons	Compost Forms in the soil over time from decomposition of brown carbon sources

adding inputs like green manure crops, crop residues, manure and compost, but not all of these inputs build stable organic matter. Some carbon sources provide a quick flush of nutrients to the next crop, while others take longer to break down in the soil and instead build stable organic matter.

Distinguishing between different types of carbon and their various roles can be confusing. I have attempted to explain it to farmers many times over the years, but I had difficulty clarifying this difference in carbon sources until an old organic farmer I work with told me his method of keeping things straight. His clear and simple way of distinguishing between the main forms of carbon found in an agricultural system and their different roles is by calling them "green carbon," "brown carbon," and "black carbon."

"Green carbon" is carbon that breaks down quickly in the soil. Its main function is to feed microorganisms, mainly bacteria, and provide a quick flush of nutrients to soil life and the next crop. Green carbon is found primarily in young living plants. Some manures are green carbon sources, especially if they are high in nutrients and low in fiber — like pig and chicken manure. Manure that contains bedding, or manure from animals that have consumed a lot of forages would not be green carbon sources (most of these would be "brown carbon" because most of the easily digested nutrients have been absorbed by the animal and what remains is a lower-nutrient, higher-fiber material). Sugar can also be considered a source of green carbon because it feeds soil bacteria, though it does not provide nutrients like young plants or manures do.

Green carbon has a low carbon to nitrogen ratio. This is important because the amount of carbon relative to the amount of nitrogen determines how the material breaks down in the soil. Because it is relatively high in nitrogen and low in lignin, soil organisms, primarily bacteria, digest green carbon quickly and release the nutrients from the green carbon source into the soil solution. Soil bacteria have a low carbon to nitrogen ratio in their own bodies, so as they die and other organisms consume them, nitrogen in a plant-available form is released into the soil. This rapid breakdown of green carbon in the soil results in a flush of available nutrients but very little buildup of organic matter. That is because some of the carbon from a green carbon source will be released back into the atmosphere as carbon dioxide when soil organisms consume the material and then respire. Only a small fraction of this carbon will remain in the soil and over time be converted into the more stable black carbon, or soil humus.

As a farmer, the important thing to remember is that if you are applying green carbon from manure or working in a young, green cover crop, the nutrients from the green carbon source will quickly become available to your growing crop. Green carbon sources break down rapidly, and go from green carbon source → to soil bacteria → to high nitrogen waste in a matter of days or weeks. This is one reason why getting sufficient nitrogen for your crop is about more than counting nitrogen units. It is possible to "grow your own nitrogen" by feeding soil bacteria a green carbon source.

On my farm I have seen a steady increase in protein content in my forages even though I do not add commercial nitrogen, and I also grow grasses like corn without adding any purchased nitrogen. One reason for this is the amount of green carbon I grow in the form of young oat, rye and legume cover crops worked into the soil in the spring. These young, green, leafy plants are a source of food for soil bacteria, including *Azotobacter*, the bacteria that turn atmospheric nitrogen into plant-available nitrogen in the soil. So even though the rye and oats aren't nitrogen-fixing crops, through feeding soil bacteria they are providing additional nitrogen and other nutrients to my crops.

"Brown carbon" is the carbon found in older, woodier plant materials. Examples of brown carbon include corn stalks, residues from small grain crops, and mature cover crops such three-foot tall ryegrass. Anything with a lot of lignified or brown plant material in it is a source of brown carbon, including manure containing bedding, or manure from animals that have consumed a lot of forages. Brown carbon has a high carbon to nitrogen ratio, which means it takes more time for soil organisms to digest it. Because soil organisms require nitrogen, and because brown carbon is relatively low in nitrogen, soil organisms must consume nitrogen from the soil as they digest brown carbon sources. This means that the amount of available nitrogen in the soil will go down for a time as brown carbon sources are digested, unless a source of nitrogen is added to the soil.

Brown carbon is consumed mainly by soil fungi. Soil fungi eat materials high in lignin and complex carbohydrates that are difficult for other organisms to break down. Fungi have lower turnover rates than soil bacteria and therefore do not provide the same rapid influx of nutrients that bacteria and their food sources do. This means that brown carbon sources do not provide a quick source of nutrients to a growing crop. The trade off is that as soil organisms consume brown carbon, more of the carbon will remain in the soil to be converted into a stable form and eventually become humus.

Crop residues are brown carbon, but they do not contribute to stable organic matter in the soil unless they break down. If you have six inches of residues lying on top of the ground, they are not being digested. It may look like you have organic matter, but undigested plant materials are not doing anything to improve your soils except for keeping a blanket on top. As the residues lay on top of the soil, they start to oxidize and much of the carbon in those residues is released into the atmosphere in the form of carbon dioxide. If you work those residues into the top layer of the soil instead, the residues will come into contact with the microorganisms that digest them. This begins the process of converting the carbon in those residues into stable organic matter in your soil. I like to work in residues in the spring, when the soil is warming up and my cover crop is growing and can capture the carbon dioxide released as microorganisms consume the residues and respire. This also provides me with a nice blend of green and brown carbon sources. The green carbon provides some nitrogen to help digest the brown carbon so I do not deplete nitrogen from my soil as those brown carbon residues are broken down.

"Black carbon" is the stable carbon that forms in soil over time as organic materials decay. It is the end product of decomposition in the soil. Black carbon is the rich black humus that provides soil with many of the beneficial properties of organic matter. Compost is partially composed of black carbon (though there is a lot of variation in the amount, depending upon the type of materials that went into making the compost and how well it was made). The black, rich-smelling portion of well-digested compost is humus.

Unlike green carbon, black carbon does not provide a quick influx of nutrients to growing plants. And unlike brown carbon, black carbon does not tie up any other soil nutrients after it is applied. Black carbon has a high CEC (cation exchange capacity), so applying a black carbon source like compost or mined humates is a way to hold nutrients in a stable form that does not leach or tie up. Applying black carbon can also improve soil structure and soil water-holding capacity, as well as providing many other benefits.

As I mentioned before, farming practices such as heavy tillage and application of ammonia cause the breakdown and loss of black carbon, or humus, from the soils. Over time this loss of humus has a negative effect on soil life, soil structure, and plant health. That is why it is very important to balance any losses of black carbon from the soil with practices that rebuild it. These practices include incorporating a brown carbon source into the soil each year, and applying black carbon such as compost or mined humates (there is more about mined humates in the next section).

Managing soil carbon in a farming system means using a variety of different carbon sources and returning plant materials to the soil at different stages of maturity. Adding both green carbon and brown carbon sources to your soils each year is a good way to balance the need for crop nutrients with the need to maintain or build stable carbon in your soils. Researchers at Michigan State University have found that growing a high biomass cover crop, such as rye, and applying compost or a brown carbon manure source (such as dairy manure or manure with a lot of bedding in it) is an excellent way to get the benefits of both green carbon and brown carbon sources. Working in the young cover crop provides the flush of nutrients that comes with bacterial digestion of green carbon sources. And by mixing that green carbon source with compost or a brown carbon manure source, you simultaneously build humus (black carbon) in your soils.

Humates

Several years ago while in Australia on a speaking tour about biological farming, I learned about mined humates. Farmers in southeastern Australia struggle to get enough water into their plants. They farm in an area of low rainfall, and their soils have very little organic matter. When I visited one of their farms I noticed a big pile of black material by the fertilizer piles. I asked what it was, and the farmer told me it was humates. He said he was applying this mined material to improve soil structure and soil water-holding capacity.

What are humates? Humates are organic material that was buried millions of years ago. They are an early step in the carbon chain that goes from organic matter, to humus, to soft coal, to hard coal, and ends with diamonds. The farther down the carbon chain you go, the harder the

The Amount of Humic Substances in Soil Organic Matter, Compost and Mined Humates

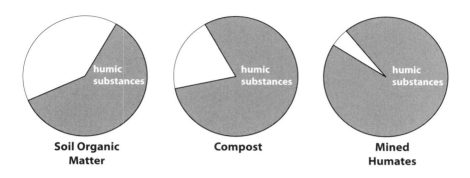

Soil Organic Matter **Compost** **Mined Humates**

Graphs show approximate amounts. Exact percentages will vary.

materials get. Soft coal, also called oxidized lignite or Leonardite, is the material used on farmland. It is found closer to the surface of the earth than hard coal, which is what we burn to run our power plants. Since it is so highly decomposed, soft coal (which I will continue to call "mined humates") is a low pH carbon source that contains few minerals.

How do mined humates differ from the humus that starts to form in your compost pile, and is the end product of organic matter decomposition in your soil? Mined humates and soil humus both contain humic substances, which are very large, highly complex molecules that result from microbial decomposition of organic materials. There may be some differences in the molecular makeup of the humic substances in the humus in soil compared to those in mined humates, but they are very, very similar. The big difference between humus in your soil and mined humates is the concentration of the material. A little more than half of the organic matter in your soil is humus, and of that humus, not all of it is humic substances. Mined humates are made up of almost pure humic substances. So when you purchase mined humates to put on your land, you are getting all the benefits of the humus in your soil, just in a concentrated dose.

Mined humates have been used as a soil amendment for many years. Because of this, a lot of research has been done around the world on the benefits of adding humates to the soil. Humates are known to:
- improve soil structure
- improve nutrient uptake
- promote plant growth
- improve water holding capacity for better drought resistance
- act as a storehouse of minerals
- make nutrients plant available

Humates improve soil structure: Humic substances in the soil help flocculate the soil and improve aggregate stability. In plain language, this means that humates help give the soil a loose, crumbly, chocolate cake-like structure.

Humates improve nutrient uptake: Mined humates have a very high CEC, which means they can loosely hold cations so those minerals do not leach. Humates also have some sites that can attract and loosely hold anions. Since very, very few anions are held by the clay/humus complex in the soil, they are prone to leaching. By adding humates to the soil, more anions (like phosphorus, sulfur and boron) are held in a form where they will not leach and are easily plant available.

Humates promote plant growth: Mined humates are commonly added to the water of a transplant solution to ease the shock of transplanting and promote plant growth. One of the consultants I work with advises tomato growers, and based on his advice a tomato farmer started adding humates to the water when he transplanted his tomatoes from the greenhouse to the field. The results were remarkable. He doubled his yields, from 25 tons of tomatoes per acre to over 50 tons per acre of tomatoes. Walking out into his field, you could literally see to the row where he put humates in the transplant solution compared to where he did not.

The tomato farmer claims the humates stimulated the tomato plant roots, but I think that is only part of the story. Humates hold on to soluble nutrients. If too much soluble fertilizer is applied to young plants, it inhibits root growth. That is because with plenty of soluble nutrients available to them, young plants won't spend their energy growing a big root system to search for more nutrients. I think one way the humates promoted root growth was by holding on to those excess soluble nutrients and thus removing them from the soil solution. The humates removed the constraint that slowed root development in the young tomato plants, allowing them to grow a large, healthy root system.

Humates improve water-holding capacity for better drought resistance: Humates improve soil structure, which helps water to soak into the ground rather than running off. A graduate student I met in Australia spent two years researching everything he could find from around the world on the benefits of humates in the soil. He found dozens and dozens of research papers, most of them from Eastern Europe, on the benefits of applying mined humates. Humates improve plant growth, nutrient uptake, soil structure and more, but none of this caught the attention of the Australian farming community. What got Australian farmers interested in using mined humates was when the graduate student proved that adding humates to the soil could improve the soil's water-holding capacity. As mentioned before, it is very dry in much of Australia, and farmers are constantly looking for ways to improve water efficiency. When the data on improved soil water-holding capacity came out, that is when Australian farmers starting applying humates.

Humates act as a storehouse of minerals: Humates are used to hold, stabilize, and enhance the uptake of minerals in the soil. Mined humates have a very high CEC (cation exchange capacity), somewhere around 100 to 300. That is close to 100 times higher than the CEC of most soils, and it means that humates can hold on to a lot of minerals. Humates also chelate minerals, which is another way of holding on to minerals in the soil.

Blending minerals with mined humates is the same idea as blending minerals with compost. Some farmers put trace minerals and a calcium source like gypsum into compost on the last turning. I saw a lot of that being done in Australia, where they have seen up to four times more trace mineral uptake by blending the traces with compost compared to putting them right on the soil. Minerals blended with compost are hooked to the organic material in the compost so they cannot leach, erode, or get tied up with other minerals so they are no longer plant available. Those minerals can then get dished out in the soil as plants need them. Mined humates have that same capability.

Both the high CEC and the chelation capacity of humates make them a storehouse for minerals in the soil. Unlike when plant nutrients are tied up with other minerals in the soil, humates do not form a tight bond with these nutrients. They loosely hold the nutrients in a form that can move easily in the soil solution where they can be accessed by growing plants.

Soybean chlorosis is a big problem for a lot of farmers in Australia. In many places the soil pH is so high, above 8.0, that it is extremely difficult to get plant-available iron. Iron applied to the soil ties up very quickly so plants cannot access it, but humates will chelate iron so it can't tie up. Because of this, many Australian farmers apply a blend of iron and mined humates to their soybeans. It is almost an instant fix. After this blend is applied, the farmers can literally see the difference in their soybean plants within days.

Humates make nutrients plant available: Mined humates are very acidic, with a pH of around 3.5. The low pH means the humates help break down naturally mined materials and minerals tied up in the soil to make them more plant available. When you mix a mineral like high calcium lime or rock phosphate with humates and then apply the blend to the soil, the humates break down the rocks and release some of the minerals. They will work the same way for nutrients that are already in the soil but are tied up in a form that is not plant available. Humates can work on those minerals to help break them down into a plant-available form.

Humates have the added benefit of being a carbon source that can hold on to minerals. So not only will humates help break down minerals into a more plant-available form, they will also hook to those minerals so they do not go directly into solution and tie up or leach.

Applying Humates

After hearing about all the great benefits of humates, you are probably wondering: How much do I need and how do I apply it? When I first learned about the benefits of humates in Australia, I went back to my farm to see if I could duplicate their great results. As mentioned in an earlier chapter, I made some big mistakes with my first load of humates. I took 1,000 pounds per acre of humates and bulk spread it on top of the ground, then I planted my crop and applied my starter fertilizer down the row. I carefully watched my crop throughout the growing season, but I did not see any noticeable difference between the field where I had bulk spread humates, and the next field over which had received manure. When I harvested in the fall, the only difference between the two fields was on one section of the field that was very sandy. In that area, yields were a little higher on the portion of that field that got humates than on the portion that got manure. Other than that, I could not see a thing from my humate experiment.

When I told the farmers in Australia about my lack of results, they told me I was applying the humates wrong. They said I needed to mix the humates with my fertilizer before spreading it so that the humates had a chance to hook to the minerals in the fertilizer. When I went back the next year and tried it again, this time mixing the humates with my fertilizer before I applied it, I saw great results. I was so impressed, I no longer apply trace minerals unless they are blended with humates. Humates make a tremendous difference in the efficiency of those traces and in the ability of my plants to access them.

I can't overstate the importance of having a lot of organic matter in the soil. Organic matter is the living organisms in my soil that are aerating it, turning over nutrients, and competing with the diseases and pests that will attack my crop. Organic matter is the recently dead organisms in my soil, including the plant residues that feed soil life and are a storehouse of minerals, and the recently dead soil organisms that release minerals and nitrogen into the soil solution. Organic matter is also the very dead, stable organic carbon that improves soil structure and loosely holds nutrients where they can be easily accessed by growing plants. As a farmer, I do everything I can to increase the amount of all of these types of organic matter in my soil.

One of the ways I work to increase organic matter in my soil is by working in crop residues, growing cover crops, and applying compost and manure. I like to apply a balance of these carbon sources because they do not all do the same thing in the soil. The green carbon sources break down quickly and provide nutrients to my next crop, but they do not build stable organic matter in the soil. In contrast, the brown carbon sources are slow to decompose, so they build stable organic matter but do not provide many nutrients to my crop in the short-term. By planting cover crops and balancing inputs of green carbon and brown carbon sources, I get the benefits of both: a quick influx of nutrients for my crop, and slower carbon decomposition that builds stable organic matter in the soil.

Applying mined humates is a way to add some of the benefits of organic matter to fertilizer. Mined humates are a complex carbon source that not only holds nutrients so they cannot leach or erode, but helps make those nutrients plant available. The low pH and high nutrient-holding capacity of humates makes them an ideal substance to blend with slow-release mined minerals, and minerals that are needed in only small quantities, like trace minerals. I have had a lot of success blending humates with calcium, phosphorus and trace minerals and getting a lot more of those minerals into plants.

By growing cover crops, working in residues, and adding manure, compost and mined humates to my soils each year, I have increased soil organic matter levels by 0.5 percent in five years. It may not seem like a lot, but given the many benefits of organic matter, small increases provide a lot of payback in terms of increased soil nutrient-holding capacity and improved soil structure and crop health.

Chapter 13

Cover Crops

Rule number four of the Six Rules of Biological Farming is about cover crops: *Create maximum plant diversity by using green manure crops and tight rotations.* I never miss an opportunity to have something growing on my land. Green plants feed soil life, build organic matter, and capture nutrients in their tissues. Keeping my nutrients in a biological cycle means those nutrients will not leach or erode, and they are in a form that is linked to biology so it is easier for plants to access them. Nutrients held in a cover crop do not show up on a soil test, but as those plants break down, the nutrients in them are released into the soil in a plant-available form. When I do not have a forage crop or row crop growing on my land, I want to have a cover crop growing. A good cover crop can provide many benefits, including:

- improving water infiltration into the soil
- reducing water loss from bare soil by evaporation
- holding soil in place and reduces erosion from wind and rain
- reducing fertilizer inputs by providing calcium, phosphorus, potassium, nitrogen and micronutrients to the following crop
- breaking up soil compaction
- producing compounds that deter weeds and crop pests
- increasing soil organic matter levels
- feeding soil biology

There are many different types of cover crops, and each provides different benefits, but there are a couple of things that all cover crops have in common: cover crops increase plant diversity in your rotation, and they pull up and hold onto soil nutrients.

Is There a Difference Between a Cover Crop and a Green Manure Crop?

I often use the terms "cover crop" and "green manure crop" interchangeably, but they are not always the same thing. A green manure crop is a type of cover crop, but not all cover crops are green manure crops.

There are four main types of cover crops: 1) those planted to build soil fertility and organic matter, and improve soil tilth — the "green manure" crops; 2) those planted to break up soil compaction; 3) those planted to manage weeds or pests; and 4) those planted to hold soil in place, improve water infiltration, and prevent erosion.

Of course, no cover crop does just one thing. They all provide multiple benefits. That is one reason why I will use the term "cover crop" to describe these plants for the rest of the chapter, regardless of their individual benefits.

The Benefits of Plant Diversity

There is a lot to be gained by adding more types of plants to your rotation through growing cover crops. Many pests, for example, prey on a relatively narrow range of species, so increasing diversity can break pest and disease cycles. Growing a wide range of plant species can also increase microbe diversity because different microbes prefer different types of plants. Another benefit of increasing plant diversity is that different types of plants access nutrients other than those the crop will pull up, so planting cover crops and working them into the soil can increase the amount and variety of plant-available nutrients. Finally, cover crops can have a different type of root system from the main crop, which will help keep channels in the soil open to allow water infiltration and air movement.

I run a lot of different test plots on my farm, and when I was first starting out I wanted to demonstrate the value of planting a diversity of plants by establishing a continuous corn plot on my farm. This sounds counterintuitive, but what made this corn-on-corn field different was that it always had one or more cover crops on it.

A lot of farmers have learned from experience that problems with diseases and pests occur when you grow corn-on-corn and don't rotate your crops. By interseeding my corn crop with clover and planting rye each fall, I wanted to show farmers that you do not have to rotate your crops to get diversity in the system — you can get diversity by adding cover crops. Planting clover and rye with my corn meant I was growing not one crop, but three. In addition, not using herbicides gave me more plant diversity in the form of weeds. Having weeds on my fields is not all bad, as long as I use management techniques like early cultivation to keep them under control.

After ten years of planting this corn/cover crop system on the same field, a University of Illinois researcher visited my farm and took a look at the field. He was very surprised by the health of the plants and the lack of crop pests. He had never seen a 10-year continuous corn field without corn rootworms in it. He could not understand how I was able to maintain such a healthy cornfield without rotating the crop. Of course, it was not really a 10-year continuous corn field, it was a corn/clover/rye field. The diversity added by the cover crops helped break pest cycles, kept the nutrients cycling, fed soil life, and improved soil structure. By adding cover crops, corn-on-corn really can be a sustainable farming system.

Increasing plant diversity results in a wider variety of soil life and insects, and as a result no one disease or crop pest can take over. If you are planting a corn/beans rotation with no variation and no cover crops, you have neither a diversity of residues nor soil life, and as a result you will not be able to stop the diseases and insects. That is one reason there are so many bioengineered crops and such heavy pesticide use: we have removed diversity from our farming system, and as a result we are in a constant battle against insects and diseases.

Plant diversity is the key that can break those pest and disease cycles. Since each type of plant uses different minerals, cover crops put different minerals back in the ground as they break down and this feeds a variety of soil life, improving the entire biological system. This means that the more plant diversity I have, the more success I have on my farm.

Nutrients Provided by a Cover Crop

Cover crops help make nutrients more accessible to your crop by pulling them out of the soil and holding them in their tissues. When the cover crop is worked back into the soil, microorganisms digest the plant material. As those bacteria, fungi, and other soil creatures die and decay, or are consumed by other soil organisms, they release nutrients into the soil in a plant-available form. They also produce proteins and increase plant-available nitrogen through their own biological processes. This means that as a cover crop breaks down in the soil, there is an increase in soil biological activity and a release of plant-available nutrients for the next crop growing on the land.

I wanted to demonstrate just how many nutrients are held in a cover crop, so several years ago I grew a cover crop and tested the plants. In early August, I planted a blend of Italian ryegrass, hairy vetch and buckwheat. Two months later, just before the first killing frost that fall, I tested stems, leaves and roots to determine what nutrients the cover crop had extracted. Based on my calculation of the number of pounds of biomass per acre produced by both the aboveground and belowground portions of the cover crop, the following table shows how many nutrients per acre were held in those plants.

Nutrients Present in Otter Creek Organic Farm Green Manure Crop

N	92 lbs./acre
P*	16 lbs./acre
K*	78 lbs./acre
Mg	16.5 lbs./acre
Ca	37 lbs./acre
S	8 lbs./acre
Trace Minerals	9.5 lbs./acre of Zn, Mn, Fe, Cu, B

*value converted to P_2O_5 and K_2O

As you can see, the cover crop pulled a lot of nutrients out of the soil and fixed some nitrogen from the air. Those nutrients will be held in the dead plant tissues and roots over the winter, and in the spring when microorganisms break down the tissues, most of those nutrients will be released back into the soil. In addition, the vetch and ryegrass will grow back in the spring and pull even more nutrients out of the soil. After I work the cover crop into the ground in the spring, those nutrients, along with carbon the plant fixed through photosynthesis, will become food for microbes and then will be released into the soil.

Working Cover Crops Into the Soil

I recommend working spring cover crops into the soil with a rotovator when the plants are young, green and succulent. If you let the plants get too large or mature, they will not provide as many nutrients to the following crop. More mature plants are slower to break down, which means it takes longer for the nutrients to be released into the soil. More mature plants can also be harder to kill with a rotovator.

I think you have greater success when you shallow incorporate your cover crop, but you can also burn down your cover crop with herbicides. Though it is not my preferred method, the benefits are still greater than the costs.

After the cover crop is worked into the soil, it is generally a good idea to wait five to seven days before planting, if soil and weather conditions allow it. If you need to plant soon after you incorporate or burn down your cover crop, be prepared to plant into heavy residues. In those cases where I need to plant into residues, I use row cleaners or a notched disc blade on my planter to make a clean seed bed where the seeds are in contact with the soil, not decaying residues. It is very important to get good soil to seed contact for uniform emergence and crop stand.

Not all plants pull the same nutrients out of the soil. Different types of plants take up varying amounts and types of nutrients. Corn and potatoes, for example, tend to need a lot of available minerals in the soil in order to get the nutrients they need. Small grains, on the other hand, can grow on low fertility soils. Oats and buckwheat in particular can grow and produce a crop on poor soils where other plants would struggle.

A researcher in North Dakota did a study looking at how well buckwheat is able to pull nutrients out of the soil. He planted buckwheat in the summer and when the plants were mature, he worked them into the soil. Several months later, the researcher went back and took soil tests from the buckwheat field and found that soil phosphorus and magnesium levels were higher even though neither of these minerals had been applied to that land. How is that possible? The soil has a reserve of nutrients that don't show up on the soil test — close to 5,000 pounds of phosphorus and 10,000 pounds of magnesium per acre — and the buckwheat plant was able to access some of them. The roots of buckwheat plants produce a mild acid that breaks down minerals that are very difficult for other plants to access, like those found in the soil reserve, or in mined materials like rock phosphate. And it is not just the major minerals that buckwheat pulls up: it can also access some types of micronutrients that we don't usually look for on a soil test. In addition, buckwheat has a shallow, dense root system that pulls in a lot of minerals and holds them until the buckwheat plant breaks down and releases those minerals back into the soil in a plant-available form.

As long as a cover crop is standing, it is holding on to nutrients. If I planted buckwheat and then went out and took a soil sample when the plants were knee high, I would not see a change on my soil test. At that point the minerals the buckwheat pulled up from the soil are still in the plant. It is not until that cover crop breaks down that the nutrients will be released in the soil. That is why a soil test is only one part of the picture — the minerals found in growing plants are the real report card of success in soil management.

I held a training class a number of years ago where I showed students a soil sample taken from a field in the spring and then another soil sample taken from that same field in the fall. Between the spring and fall soil samples, I had applied a ton of Bio-Cal to the field and then planted sweet clover, yet the spring and fall soil samples looked nearly identical. I told the students that if I was the farmer who had purchased that Bio-Cal I would want to know why my soil tests did not show any increase in calcium. What did that Bio-Cal do for me, I asked the students. The students said they would tell the farmer that there were no changes on the soil test because it takes time. That may be true, but there is more to it than that. I then showed them a tissue test taken from the roots, stems and leaves of the sweet clover that was growing on the field. It had really high calcium levels. That is where all the Bio-Cal went! Nutrients held in a cover crop are in a form that cannot leach, cannot erode, and cannot get away. As the sweet clover cover crop breaks down, much of that calcium, along with the other nutrients in the plants, will get released into the soil for soil biology or the next crop to use.

Freshly rotovated soil–building cover crop.

Using Cover Crops to Build Soil

I like to give land a "rest" every few years and rather than grow a crop, spend time soil building. All land would benefit from this, and because I farm organically, a perfect opportunity for soil building is during the years of transitioning fields from conventional to organic production. During the two growing seasons of the transition period, the land has to be farmed organically but whatever is produced from that land has to be sold on the conventional market. This means that I can put crops back into the ground without taking as much of an economic loss as I would have if the land was at full-scale production.

One farm where I spent time building the soils is a few miles down the road from our home farm. The land is owned by a couple from the city who like to come out to their country home to enjoy the quiet and solitude, and to go hiking and cross country skiing on the trails they have created along the field edges and through the woods. They used to rent the crop land to a conventional farmer who planted corn and soybeans and used conventional fertilizer, herbicides and insecticides, but they asked the renter to leave after he sprayed herbicide on the beans right before a heavy rain, and the combination of erosion and herbicide runoff created ruts through their hiking path and killed all the grass. Whatever that herbicide was, they were unable to get grass reestablished on their walking paths. They were really upset, so they called and asked us to farm their land organically. I said we would, but on the condition that we could have the land rent-free for the two years it would take to transition it to organic.

The owners agreed, so in the fall after the conventional farmer harvested his bean crop we took soil tests and added high calcium lime and rock phosphate soil correctives. The following spring the fields were very weedy, so we rotovated down the weeds and planted one bushel of buckwheat per acre with ten pounds of red clover. After that we walked away for the rest of the growing season. Later that summer, the owners of the farm called me to say that the buckwheat was in bloom, and the field was solid white and noisy from all of the honey bees.

I let the buckwheat go to seed that fall and since clover is a perennial, the following spring I had a second crop of clover and buckwheat on the fields. A lot of farmers think it is crazy letting buckwheat go to seed because it is hard to get rid of once it is established. Having volunteer buckwheat in a field does not bother me because I have rotovators and cultivators to work it in, and the little that does survive in the rows does not compete that much with my crop. In the meantime, I get a second cover crop from my one buckwheat planting.

The second year of transition to organic I turned down the buckwheat and clover in the summer and planted Italian ryegrass. I left the ryegrass on the field that fall, and worked it down when it came back in the spring, which was the start of the third growing season I had managed the land. The land was now certified organic. I did not harvest any crops for two growing seasons while we were going through the transition to organic production, but I also spent very little money on those fields. And after two years of soil building, I knew that soil would be loose and crumbly and full of life. The fields were primed with nutrients and organic matter, and I was set to grow my most profitable crop with the greatest opportunity for success.

Cover Crops I Like to Plant

On my farm I grow a wide variety of cover crops. In part I do this because I like to keep the level of plant diversity on my farm high, but also because some cover crops work better than others for different situations. For example, if I am planting a cover crop after corn, I generally plant cereal rye grain because it is easy to establish, even late in the year, and will come up again early in the spring, holding nutrients and provide biomass for me to work into the ground before I plant. (One caveat about cereal rye grain: this plant forms a massive root system, so without the right kind of equipment and know-how, it can be hard to kill.)

If I am planting a cover crop after I harvest soybeans in the fall, I won't plant another legume because I am not after more nitrogen. Instead, I will plant oilseed radish to break up compaction or a late-season cover crop like cereal rye grain. If I do not get a cover crop planted in the fall, I will plant oats the following spring. Growing oats is one of my favorite spring cover crops because it is very good at accessing nutrients, which means it does fine on hard, tight soils without a lot of life in them. Planting oats is also a great way to hold nutrients after applying manure in the spring. Instead of the soluble nutrients from the manure staying in the soil to fertilize a healthy crop of weeds, those nutrients are taken up by the oat plants so there are a lot fewer left in the soil for the weeds. Those nutrients will be tied up in the oat cover crop for a time, but I know I will be able to access them later when I work the cover crop back into the ground and it breaks down in the soil.

I often talk about the benefits of planting a spring oats cover crop at my meetings, and a few years back some farmers from Minnesota decided they would follow my advice. That winter, after I had finished speaking at one of my meetings, they came up and told me that they had planted 10 bushels of oats per acre in the spring, just like I said. I know I talk fast, and apparently they missed a few details. I did not say to plant 10 bushels of oats per acre, I said don't spend over 10 dollars per acre on oat seed! That is closer to three bushels of oats an acre than to the ten bushels they planted. Even though they got my advice mixed up, they were just ecstatic with the results. They said the oats came up just as thick as hair on a dog — it looked like a lawn. Once it got knee high they rotovated it down and then planted beans. They said the bean crop was weed free and the highest quality, best looking beans they have ever had.

Buckwheat is another cover crop that works well on my farm. I have already mentioned buckwheat's ability to pull nutrients out of the soil, but its other great attribute is its ability to suppress giant ragweed. It is really hard to get rid of giant ragweed when you farm organically because you can't rotary hoe it out. About the only thing to do is to hand pull it. It is probably the biggest weed challenge for organic farmers here in the Midwest. Several years ago the ragweed on one of my fields started to get out of control. We hand pulled it; we hired neighbor kids to help pull it; we mowed all the headlands and all the grass waterways so the ragweed wouldn't go to seed. It was a huge battle. Last year, I pulled every giant ragweed myself in a half hour. What changed? I planted buckwheat.

A farmer I work with in Ohio had a similar experience with buckwheat and giant ragweed. He had a big problem with giant ragweed on one of his fields so he planted buckwheat. He covered the whole field except for one corner where he planted sweet corn for his own use. He rotovated in the buckwheat that fall and the following spring he planted soybeans. That season there was not a single ragweed on the entire field except for the corner where he had planted his sweet corn — and there it was solid giant ragweed. That was pretty good evidence to him that buckwheat really does suppress giant ragweed. In just one year, the part of the field that had buckwheat planted on it was free from giant ragweed.

I don't want you to get the wrong impression and think that buckwheat will solve all your giant ragweed problems. In a farming system there is always a lot more to solving a problem than doing just one thing. But on my farm and the Ohio farm, whether we hit it just right with timing of working in the buckwheat, or the weather was just right, or some other combination of factors, the buckwheat sure did make a huge difference in controlling giant ragweed. On those fields, what we saw was real. How buckwheat works to suppress ragweed, and whether it will work on all farms, is a good subject for a research project somewhere.

The Right Cover Crop for Your Farm

The benefits of cover crops hold true no matter where you live, but which cover crops grow best and fit your farming system will vary a lot from place to place. In southern Wisconsin where I farm the length of our growing season can present obstacles. Some years it is difficult to get a cover crop established in the fall after my corn or bean crop comes off. Farther south, farmers have a longer growing season and more opportunity to plant cover crops. Some farmers I work with in Missouri, for example, use a highboy to plant hairy vetch and Italian ryegrass into their corn the last week of August, and later that fall they have a nice cover crop established. They have had good success with this method. That works well for their farming system in their part of the world, but it does not work for me in Wisconsin because my growing season is shorter.

Interseeding a cover crop into an existing crop is an interesting idea that I think needs more research. In most corn/bean farming systems, by the time the crop is harvested it is too late to get most cover crops established other than cereal rye. Another option would be using a highboy or flying on seed to plant into a mature growing crop in late summer or early fall. Ryegrass, clover, oilseed radish, tiller radish and oats are all crops that could work in this type of a system. Tiller radish has been getting a lot of attention lately as a good cover crop for breaking up soil compaction, but so far attempts at interseeding have only been successful about 70 percent of the time. I am optimistic that as more research and more trials are conducted on interseeding, germination rates, and crop establishment, success with this system will improve.

Another way to get more diversity into your rotation is by adding forage blends for livestock feed, a system that works really well on organic farms. It gives the soil a one or two year break from cropping and eliminates fresh weed seeds. This is one reason why organic farming and livestock fit together so well.

All cover crops have different attributes, and it is important that you choose the right cover crop for your farm and your needs. For example, legume cover crops will provide nitrogen but produce less biomass than other crops. Oilseed radish can help break up soil compaction and will suppress certain types of nematodes, but it will not provide a lot of ground cover and will winterkill. Cereal rye grain will overwinter and provides a lot of biomass, but it is extremely important to work this crop down in the spring before it gets too mature or lignified, and even then it can be difficult to kill. These are just a few of the dozens of different cover crop plants, all with different attributes, and some that work better in different farming systems or different parts of the country than others. Farmers should do some research or talk to a consultant before choosing a cover crop for your farm. There is also an excellent resource on cover crops available through the Sustainable Agriculture Network titled *Managing Cover Crops Profitably* that can get you started. But whatever cover crop you choose, remember: the more diversity you have, the better!

The following table highlights some of the benefits of cover crops commonly grown in the upper Midwest. The numbers under the headings on the table indicate how strongly that crop exhibits each attribute, with 0 being "poor," and 4 being "excellent."

The farmer who won the conservation award for Australia several years ago did it by growing cover crops. Most farmers in the dry part of that country leave the ground bare between crops. They do not want to plant cover crops because they only get 12 inches of rain a year, and they are afraid that the cover crop will use too much water and then their crops will fail. The farmer who won the conservation award went against this practice

Attributes of Key Cover Crops

	Planting Rate (lb)	Planting Depth (in)	Alleopathic Effect	Chokes Weeds	Loosens Subsoil	Loosens Top Soil	Frees Up P and K
Italian Ryegrass	5-10	⅛ to ½	2	4	2	4	2
Winter Rye	60-120	¾ to 2	4	4	1	4	3
Oats	80-110	½ to 2	3	4	0	3	3
Buckwheat	50-70	½ to 1½	3	4	0	3	4
Sorghum-Sudangrass	35	½ to 1½	4	4	4	2	2
Hairy Vetch	15-20	½ to 1½	2	3	2	3	2
Red Clover	8-10	¼ to ½	2	2	3	2	3
Radish	8-13	¼ to ½	3	4	4	2	3
Mustard	5-12	¼ to ½	3	3	1	3	2

Scale 0-4 with 0=Poor and 4=Excellent

rops, burned them down with herbicide so he could
dues on top of the ground, and built his own drill for
h those residues. What this farmer realized is that
on top of the ground and get roots established to
a actually have more water for your crop. Leaving
allows the water to evaporate from the surface of the soil,
soil structure deteriorates it becomes more difficult for water to soak
in. Rather than reducing the amount of water available for the crop, the
cover crop improved water infiltration and improved soil quality as well.

Last year I planted 400 acres of corn and beans, and every one of those
acres had a cover crop. I am always working to capture and hold nutrients,
and I also want a blanket on top of the ground to protect the soil life from
freezing and from high heat. I do everything I can to promote healthy
soil life on my farm, and planting cover crops is an essential part of the
system. Cover crops also provide other benefits, such as breaking pest
cycles, suppressing weeds, improving water infiltration, and breaking up
soil compaction. Finding the right cover crops for your farm can take some
research, not to mention some trial and error. But once you find the ones
that work well in your farming system, the benefits more than make up for
the cost of the seed and a trip across the field.

Of all the inputs you use on your farm, seeds are the cheapest thing
to transport and spread. Along with addressing more minerals than just
nitrogen, phosphorus and potassium (NPK), growing as many different
kinds of plants as you can, as often as you can, is the greatest improvement
you can make on your farm. Cover crops have the potential to reduce
chemical use, reduce the need for applied nitrogen, and improve soil
structure and soil biology. I believe that the greatest opportunity for success
on any farm comes from this system, but you need to be committed to
it and invested in it. When you are, you will have the potential to grow
your highest return crops, year in and year out, more sustainably and more
profitably.

Rye cover crop with dense root system.

Chapter 14

Tillage

Rule number five of the Six Rules of Biological Farming is about tillage. *Use tillage to control the decay of organic materials and to control soil air and water. Zone tillage, shallow incorporation of residues and deep tillage work great on many farms.* As a farmer I have jobs to do, and two of those jobs are to manage decay of residues and to control soil air and water. Under ideal conditions it is possible to do these jobs without tillage, but who farms under ideal conditions? We plant when the ground is wet, we harvest when the ground is wet, we drive heavy equipment across our fields, and all of these things pack down the soil. Do we just stand back and hope the soil compaction gets better on its own? I think we need to till to work in residues, to break up surface compaction and keep the soil breathing, and to break up subsoil compaction, but this does not necessarily mean plowing, chiseling, or more trips across the field. Tillage should be *thoughtful disturbance of the land.* Just because my neighbor is plowing and it is Monday morning that does not justify me going out to plow. I till only when I need to till to get a job done.

Why I Believe Some Tillage is Necessary

Like any living organism, the soil needs to breathe. Without air movement in and out of the soil, the soil biology will slow down and many more anaerobic areas (areas without oxygen) will develop in the soil. This is a problem because most of the beneficial soil organisms need oxygen, while denitrifying bacteria (the bacteria that cause nitrogen to be lost from the soil) thrive in an anaerobic environment. Our objective is healthy, biologically active soil that is breathing; taking in oxygen and giving off carbon dioxide. If you do not till, there is the potential for a crust to form on the soil's surface, which makes it difficult for the soil to breathe. Shallow tillage breaks up that crust, allowing soil life to get the oxygen it needs.

Tillage is also needed to break up subsoil compaction so water and roots can penetrate deeper into the soil. Driving on the soil when it is wet is the biggest cause of soil compaction. That is why I say, "Any condition at planting less than ideal is unacceptable. Until June 1st." At some point you need to get a crop in, which means sometimes we end up planting in four-wheel drive and mudding crops in. Driving on fields when they are too wet packs down the soil, and if I pack it down, I am going to unpack it. That's why I subsoil. When the soil gets compacted, I run my deep aeration tools: either a zone tool or a plow subsoiler. I do not want to rip up the soil, but I do want to make slots for water to move down, and I want to fracture any hardpan so I can grow deep roots and prevent the soil from becoming waterlogged. Once I get roots and more organic matter, I have more soil life and a deeper aerobic zone and I continue to improve my soil structure. And as the soil structure improves, I have less need to till, and I won't need to run tillage tools to aerate the soil as often.

Roots Penetrating into the Subsoil

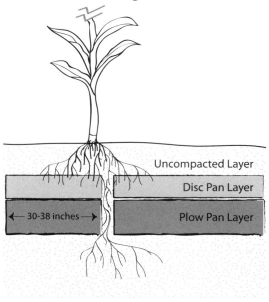

Uncompacted Layer

Disc Pan Layer

30-38 inches

Plow Pan Layer

Another reason to till is to speed the breakdown of residues. When residues sit on top of the ground, much of the carbon in those residues is oxidized and lost to the atmosphere, rather than being consumed by soil organisms and returned to the soil. The soil also takes longer to warm up in the spring underneath a mat of undecomposed residues. Shallow incorporating residues increases contact of the

Subsoiler.

residues with soil organisms, which speeds decomposition, allows the soil to warm up, and feeds soil life. In addition, having some residues loosely mixed in to the top layer of soil allows them to act as wicks for moisture and helps prevent a surface crust from forming. More farmers are starting to adopt this method of residue incorporation, sometimes called vertical tillage.

Churning up the top several inches of soil to incorporate residues may seem like invasive tillage, but while you can hurt soils when you till them when the ground is bare, you do not hurt soils when you till in a lot of plant material or residues. I demonstrate this at our field days each summer by rotovating down six-foot tall green corn plants. It is pretty difficult to shallow incorporate that much plant material, so instead, I work it into the top six to eight inches of soil. By working it in that deep, I put as much of that corn residue into contact with soil life as possible. Far from harming the soil, this feeds soil life, builds organic matter and deepens the soil's aerobic zone.

One year, a farmer stopped in about a week after our field day and asked what the corn residues looked like now that seven days had passed. We went out to the field to take a look and saw that while there were still residues on top of the soil, when we dug down into the soil, the majority of that corn was rotted down. The corn was still green when I worked it in, so putting that succulent plant material into contact with soil life meant it broke down quickly. Older, lignified residues will take longer to break down, but working them into the soil is the best way to speed that process.

Another reason to till is to manage carbon dioxide, which plants need to grow. Using sunlight for energy, plants convert carbon dioxide from the air into the sugars and carbohydrates that make up their biomass. The amount of carbon dioxide available to plants is limited somewhat by what is in our atmosphere, but another source of carbon dioxide is from the soil. Most soil organisms, just like animals, take in oxygen and give off carbon dioxide. When you till you introduce more oxygen into the soil and this causes an increase in the activity of microorganisms, which results in an increase in the amount of carbon dioxide that is released. If you till when the soil is bare, this carbon dioxide is lost to the atmosphere. But if you till when there are plants growing, the extra carbon dioxide released by soil organisms can be taken up by the plants and results in a boost in their growth.

Many farmers have observed this effect firsthand after cultivating in the summertime. Soon after cultivating, the crop will be noticeably greener and taller. What causes that? Putting more air into the soil increases biological activity, which causes an increase in the amount of carbon dioxide released from the soil, allowing the plants to take up more carbon dioxide and resulting in that observable boost in growth. I believe that if you really want to grow 500 bushels per acre of corn, timing tillage and releasing extra carbon dioxide when your crop is growing is an essential piece of the puzzle.

Problems With Too Much Tillage

While some tillage is a good and necessary thing, more tillage does not bring more benefits. Tillage should be done only when there are jobs to do, like breaking up compaction or speeding the breakdown of residues. Remember, your tractor doesn't need exercise! I don't want to plow, dig and disc, and I think a chisel plow is just as bad as a moldboard plow at tearing up soil aggregates and disturbing soil life. Too much tillage degrades soils because it causes a loss of carbon (organic matter) and it damages soil life.

One of the problems with aggressive tillage is that it introduces a lot of air into the soil. As already mentioned, when oxygen is mixed into the soil through tillage, there is a boost in the activity of microorganisms. Those microorganisms consume soil carbon, organic matter, and give off much of that carbon in the form of carbon dioxide. If there are plants growing on the soil to take up that carbon dioxide, it is held in the growing plants and can be returned to the soil again later in the form of residues. But if the soil is tilled when it is bare, for example chisel plowing in the early spring before planting, there are no plants there to capture that carbon dioxide and it is lost to the atmosphere. This is why over time aggressive tillage results in a loss of soil carbon. If you avoid tilling the soil when it is bare, you can slow this loss of carbon from the soil.

This is one reason why no-till is promoted as a means of sequestering carbon in the soil. Without tillage, it is believed that the activity of microorganisms will be in balance with plant growth and residue decay, and soil carbon will build over time. Unfortunately, there is a lot of conflicting research on this topic, so the jury is still out on whether or not no-till farming increases soil carbon. I am sure that under some conditions it does increase carbon sequestration in the soil, but in many situations it does not. As long as large quantities of nitrogen are put on crops, it will be really difficult to sequester carbon. In my opinion, a better way to increase soil carbon is to grow cover crops, apply compost and humates, shallow incorporate residues, and till only when necessary.

Another problem with aggressive tillage is that it disturbs soil life and the root zone. When you turn over the soil with an aggressive tillage tool like a moldboard plow or a chisel plow, the top six to eight inches of soil gets mixed and turned. When this happens, the fungal networks that give soil its loose crumbly texture are ripped apart, soil aggregates are destroyed, root channels and earthworm channels are broken up, and mycorrhizal networks are ripped apart. We need soil life like mycorrhizae to supply plant-available nutrients (mycorrhizae bring available phosphorus to the plant) and we need soil aggregates to keep the soil loose with plenty of air and water infiltration. I also do not want to tear up the root zone with tillage.

The highest level of biological activity in the ground is in the rhizosphere, the area right around the roots. That root zone is full of moisture and carbohydrates given off by plants to feed soil life, and therefore has a really high level of biological activity that I do not want to disturb.

The most fertile ground in your soil is around earthworm channels and plant roots. Those four to 12 inches of "middle zone" where most of the plant roots are found is what I want to establish and leave alone. If I dig up that zone and turn the soil over, soil life has to start over. Earthworms will need to dig new channels, roots will need to find new paths to grow in because their old ones have been torn up, soil fungi will have to regrow, and soil aggregates will have to reform. This takes time, and until soil life can get reestablished, soil structure and soil biological activity will suffer. Subsoiling (cutting a deep, narrow slot into the ground with tines 30 to 38 inches apart) does minimal damage to the middle zone. The same is true of shallow incorporating. But once you decide to do major aggressive tillage in the middle zone, you have decided to start over. It takes a long time to rebuild what tillage tears up, so you should be sure that you have a good reason before doing it.

Even though aggressive tillage has a lot of negatives, in some cases I think a chisel plow is the right tool to get the job done. I will on occasion use a chisel plow or other deep mixing tool in the fall to incorporate a lot of manure or soil correctives into the soil. I only use the chisel plow when I have a lot of inputs to mix in, and only in the fall when there are plants or residues on the ground. It is certainly not a tool I would use every year, but under the right conditions it does have its place.

One fall I encouraged my son to use a chisel plow to break up compaction and work in a large application of manure and fall soil correctives. I can guarantee you he will never do that again. My son's home farm is on a ridge, and a road divides the main ridge-top field into two halves. One year after he harvested his last hay crop he put on a coat of manure (yard scrapings) and his calcium soil corrective on one half of his ridge-top field. He needed to work in the manure and soil correctives, so went out and chisel plowed it. Why did he chisel plow? Because he had a lot of inputs to work into the soil, and because our hay fields get really compacted. Between the heavy baling equipment from making big round bales, the 12-bale trailer that we drive across the field to pick up the bales and take them to the barn, and the weight of the truck and the trailer, a lot of compaction takes place. Hay fields also get a lot of traffic because we make four or five cuttings each year, and more than once we had to drive on the field when it was wet.

So my son decided to go in and chisel plow the fields in the fall to work in those inputs and break up some of the compaction before he planted the field to corn the next year. He had originally planned on doing this for the entire ridge-top field, but he ran out of manure before he got to the second half of the field, so he decided he would wait to plow that half until spring. So half of the field got manure and soil correctives mixed in with the chisel plow, and the other half of the field got a fall calcium soil corrective and no tillage. The following spring both fields had alfalfa growing on them — you can't kill alfalfa by chisel-plowing with a narrow shovel — so he rotovated both fields to kill the alfalfa. He then spread manure on the half of the field that needed it and planted his corn. That summer was quite dry, and the corn on the field that was chisel-plowed the previous fall suffered. By late summer it was brown and burnt up, while the field that did not get chisel-plowed was lush and green. Later, when we harvested the corn, there was a 35 bushel per acre yield difference from one side of the road to the other. That chisel-plowed ground just wiped us out. Why?

I have my theories as to why the chisel plowed field suffered, but one year is not a trial, and I doubt we will ever get a chance to test this again. That chisel plow will sit in the weeds and the tires will rot off of it long before my son ever hooks his tractor on to it again. He is convinced the chisel plow cost him 35 bushels of corn per acre. I told him that when you have a lot of lime and a lot of manure, you have got to work it in to the ground somehow, but he has decided we will be shallow incorporating from now on and running a big ripper to break up compaction and assure drainage and deep root development. Whatever happened on that chisel-plowed field, whether we tore up the root channels, destroyed the earthworm burrows, burnt off a lot of carbon, or just had a lot less green manure alfalfa to work back into the soil in the spring, the difference between that field and the rotovated field was really noticeable. It was only one field on one year and maybe that will never happen again, but my son is not about to find out!

No-till

Many farmers choose no-till because they believe it reduces erosion, increases carbon sequestration in the soil, and protects earthworm channels, complex fungal networks and other important soil life. But there are problems with no-till. How do you work in your soil correctives if you don't till? If your soil has a 5.5 pH and you need to apply lime, how do you get that lime into the soil? And how do you address compaction without tilling? Surface compaction causes a crust to form on the soil, which means the soil can't breathe and water can't infiltrate, and subsoil compaction forms a layer that roots can't penetrate. One of the main reasons you till is to manage soil air and water, so unless you have high-organic matter soils, with tons of residues being digested, and lots of earthworms working and mixing and aerating soils, you had better do something to get air and water into your soil. Remember, you till for a reason, and you need to earn the right not to till.

Every no-till farmer I know says that soil structure will get better once earthworms get established, which can take about five years. Earthworms do help break up surface compaction, but even with earthworms it can be very difficult to break up a deep compaction layer. When you have subsoil compaction, your roots cannot grow deeper into the soil to access water and nutrients. I see a lot of tight soils and flat, stunted roots on no-till fields. You also won't find a lot of earthworms on compacted, waterlogged soils because the worms need air. It is critical to break up compaction and get air and water down in that soil, and that is why sometimes you need to till.

Now don't get me wrong, I am not opposed to no-till farming, I just do not think it works in every situation. Only about 20 percent of the farmland out there is suitable for no-till. Well-drained, loamy soil with six percent or more organic matter and a good balance of minerals will naturally have good soil structure, and the problems with compaction and air and water management usually encountered in no-till will be less likely to occur there. The rest of the farmland needs intervention. On most farms, at some point during the growing season we end up driving on the fields in wet conditions and packing down the soils. If the soils are compacted and you are no-till, what are you going to do to get air and water back into the soil? You certainly have to earn the right to reduce or stop tillage, and there are always situations where some form of tillage is necessary.

It is also very difficult to incorporate nutrients without tillage. When you put soil correctives or fertilizer on the top of the soil, much of that material will stay there or take a long time to work its way into the soil. If you do not till there is also the potential for surface-applied manure to run off in a heavy rainfall. Even if runoff does not occur, the nutrients from the fertilizer or manure will remain in just the top few inches of soil. The top inch of soil on a no-till field has all the toxins from pesticide application, and a high concentration of fertilizer nutrients. Plant roots, however, are not concentrated in just the top couple inches of soil, which is one reason why I want those nutrients to be down deeper where the soil roots extend, pulling up nutrients. Another reason to get nutrients deeper into the soil is because the top layer of soil dries out first. Nutrients need to be in solution in order for plants to access them, so if the nutrients are all in that top, dried out layer of soil, they are not available.

Zone Tillage

Zone tillage is a type of tillage that is between conventional tillage and no-till. Farmers who practice zone tillage plant and till in one pass and in one zone so they do not tear up the whole field. They subsoil to allow roots to grow deep in the soil, and they also put their fertilizers and nitrogen deep in that zone where roots can access them as they grow, which means the plant roots reach that deep nitrogen later in the growing season.

Their goal is to reduce the amount of tillage they do, and to reduce the number of trips across the field. They also want to create an area with a higher concentration of minerals deeper in the soil that is friendly to root growth. The late Don Schriefer, author of *From the Soil Up* and *Agriculture in Transition*, was a big proponent of zone tillage.

No-till is often touted as a way to prevent erosion, but I have seen many examples of no-till fields that had much worse erosion than nearby fields where the farmer grew cover crops and tilled the land. On one biological farm I work with, they got six inches of rain in a big storm one spring. The local USDA soil conservation person was very upset that the biological farmer was no longer practicing no-till, and was concerned that the recent storm had caused a lot of runoff from his farm. So the farmer told the USDA employee to come out to his farm and take a look. On one side of the road was the recently planted no-till farm. That field had pools of standing water and big gullies from erosion and runoff. On the other side of the road was the biological farm, which had been subsoiled in the fall and now had a freshly cut cover crop laying on top of the ground, with not a gully or puddle in sight. The difference does not come down to till versus no-till as much as healthy, beautiful soil structure compared to hard, tight, compacted soils.

Just because you till does not necessarily mean your soils are going to be more degraded and you will have more problems with erosion than farmers who choose not to till. Tillage can actually improve soils when it is done correctly and when it is done for a specific purpose: "thoughtful tillage," as I like to say.

What I Do On My Farm

Every field on my farm has something growing on it throughout the year. If a field is not in pasture or producing a forage crop, I plant cover crops after I harvest or interseed into my standing corn crop so there is always something growing. That is one reason why the rotovator is one of my most important tillage tools. Each spring I use it to work the cover crops in to the top three to four inches of soil. I also like to use the rotovator to break up old sod or work in an alfalfa field that I will be planting to corn. When I first started using the rotovator, I thought it made hamburgers out of earthworms, but I later realized that only happens when you go too deep, rotovate at the wrong time, or work bare soils.

I run a 12-foot wide rotovator, and farmers often ask me why I don't get a bigger one. They seem to think that it takes too long to work in a cover crop or old alfalfa with such a narrow piece of equipment. I agree that the rotovator is slow, and it takes a lot of horsepower to run it. But I can also kill alfalfa in one pass. Most farmers kill alfalfa by discing it twice and then running the soil finisher. If you have a 30-foot disc and a 30-foot soil finisher and you make a total of 3 trips, and I have a 12-foot rotovator and I make one trip, who got more done today? And I drove over the soil fewer times. I also like having the blanket of plant residues on top of the ground that the rotovator leaves behind.

Rotovator blades out of the soil.

I like to plant my corn in 38-inch rows. Most farmers plant their corn in narrower rows, and after years of hearing all the arguments for planting on narrower rows I decided maybe I ought to try it. So a couple years ago I went out and bought a 30-inch corn planter. I plant cover crops, and in the spring after the cover crop is rotovated and all the residues are starting to break down, we go in and plant. Why didn't anyone tell me what a mess it would be trying to plant in 30-inch rows through heavy residues? The problem is, you can't plant in that much residue, you have to plant through it. We have got a notched disc blade-type row cleaner on our planter, which has worked well for moving aside residues when planting on 38-inch rows. Well, the year we were trying out the 30-inch row planter, due to the weather we had to plant around the clock for a few days, and I got the 10 p.m. to 2 a.m. shift. I was out there planting at night, pulling a six-row planter with a new 125 horsepower tractor, when all of a sudden the tractor started idling down. I thought, "Oh no, something's wrong." I stopped the tractor and went out and looked and saw that I had turned my planter into a snowplow. The residues were so thick I could not get the planter through them. The planter was pushing a huge pile of residues, and a drive chain had gotten wrapped with residues and broke. It was the middle of the night, and I was standing out there in that field in the pitch black wondering what to do next. I just shut off the tractor and went home.

The next day I had to decide how to go forward. All the residues were still on the field, and I still had that 30-inch row planter. Someone suggested I wait a week and start over — till the residues in once more, then try planting again. The problem with that idea is if I till again with nothing growing on the land, I will burn up all of my carbon. Instead, I fixed the planter, raised the row cleaners up, and continued planting. The planter did not break this time, but guess what? The crop didn't come up. At least, not very well. The field was thin and patchy, not uniform like my usual picket fence corn. Education is never cheap! In the end, I just worked everything back into the soil as a green manure crop.

But that was not the worst of it. On the fields that did get successfully planted on the 30-inch rows, a few days later it was time to go out to rotary hoe. I don't know how anyone does that! My knuckles were white from hanging on to the steering wheel, trying to get that tractor between those narrow rows, going around curves on hillsides. There is no room out there! Then the rotary hoe plugged up. Let me tell you, we sold that 30-inch planter real quick and went back to our 38-inch rows.

The next year we were back to our usual system of planting on 38-inch rows in furrows. We can grow 200-plus bushels of corn per acre on 38-inch rows, so I do not believe planting more seed in narrower rows is our limiting factor. Success on my farm is about managing the whole system.

My farming system includes shallow incorporating cover crops, keeping residues on the field, and putting compost on the top of the ground in spring. This means there are a lot of residues on the ground in the spring, and I have found that the best way to plant through all of that residue is to plant in furrows. I use a notched disc blade and row cleaner to create a clean seedbed with good soil/seed contact while also keeping all of my residues on the ground. I want to feed my soil life and keep something on the soil at all times, and the furrow planting system works best for me. It is also a lot easier burying weeds when you start with furrows, rather than starting out level and piling soil up in the row.

Three days after we plant we go out and rotary hoe. At this point, the weeds are small, what I call in the "white" stage. This is the best time to get

Planting in furrows.

Rotary hoe.

out in the field because it is easiest to control the weeds at that size. The rotary hoe also helps to keep the surface of the soil dry, loose and crumbly. When the corn gets three to four inches tall we go in and rotary hoe a second time, wiping out more young weeds and keeping the top dry.

I like the rotary hoe because it is fast, cheap and effective. With a 30-foot machine, I can cover about 30 acres in one hour.

When the corn gets eight inches to a foot tall we go in and cultivate. One of the neighbors stopped by not too long ago and asked what kind of cultivator I have. I told him I have a yellow one, a Buffalo cultivator. He said, "I went by your field and it looked like a hay field there were so many weeds in there. Then I went by again the next day and there wasn't a weed left. How do you do that?" I told him that when the corn is a foot tall and the weeds are about four to six inches tall I go out and cultivate and bury the weeds. In order to be successful at this stage, the corn does need to be taller than the weeds. But an advantage to working in larger weeds is that they have enough biomass that they act like a green manure crop.

Because we plant in furrows, two trips with the rotary hoe and two cultivations each year cleans the field right up and we grow a good corn crop.

Buffalo used to say it was not possible to plug their cultivators. Now they say that most of the time you can't plug their cultivators. I think I was the first person ever to plug up a Buffalo cultivator. It happened because we have so many residues. Someone once suggested that we cultivate at a different time, when there were not so many plants growing and so much residue on the ground, but why would I want to burn off my carbon by cultivating at the wrong time? It is much better to work the soil when there are plants growing there that can utilize that carbon dioxide for their growth.

Now, if I was a conventional farmer, I would do things a little differently. I would work down my cover crop, then plant in fresh clean soil and band my herbicide. I would then come back and make one cultivation after the crop was up. If I banded herbicides down the row and had my residues in the middle, where would the weeds grow? There would be no place for the weeds to get established. By leaving residues in the row and making one cultivation when the crop is growing, I also release carbon dioxide at the right time — when the crop can utilize it and get a bump in growth. I also

Buffalo cultivator.

think it would work well to seed a green manure crop like clover, rye or a brassica into the corn at last cultivation when the corn is knee high, but I do not know of many farmers who do this, so it is an idea that needs more research.

Just like too much tillage can burn off soil carbon and disturb soil life, too little tillage can result in cold, tight soils with little air and water movement and a layer of undecayed residues on top. Tillage is not something we should do because that is what the neighbors do, but we should not avoid tillage because that is what the neighbors do, either. Every farmer needs to evaluate his or her own soil and crops to determine how much and what type of tillage works best on that farm.

The number one rule of any tillage system is that it should not place limitations on yield. If you are chisel plowing every spring, or practicing no-till, and you see problems with loss of organic matter, soil compaction, erosion, or if you need to use more and more chemicals each year, you need to reevaluate your tillage system. I call tillage "thoughtful disturbance of the land" because it is not something to be done without forethought.

Tillage is another tool to help you make things better on your farm, and how much and what type of tillage to do is something you should be continually evaluating for your farming system.

Chapter 15

Final Thoughts

As I finish this book, I am sure I have missed some things, and I am sure everything is not explained in enough detail for everyone. I am equally sure that some things are not 100 percent agreeable to everyone. But please remember, this story is not about all of those details; it is about seeing the big picture. Have you taken care of soil biology? Have you done all you can to create an ideal home for soil life? Are the soil organisms being fed? Do they have the minerals they need in abundant amounts?

After all, there are many roads to Rome, but to get there you do have to know where you are going.

This book is a collection of my stories, my observations, and my studies. It is the things I have witnessed on a number of successful farms with beautifully healthy soils and incredible crops. True, some things on these farms could be happening by accident, but some things were part of someone's masterful plan, and one thing is sure — none of the successful farms violated the principles of biological farming.

The opportunity to feed people quality food while improving our air, water and food nutritional value is not only achievable, but it is being achieved.

It is not about who is right, it is about what is right. The information in this book is not mine or my daughter, Leilani's — it is an overview taken from a large base of testing and knowledge. This book covers much of the information I have gathered over a lifetime of seeing, doing, studying, testing and consulting.

With this book, I hope to get more farmers understanding and practicing biological farming. I wish it were more black and white, faster to fix, and that everything always worked perfectly right away, but it is not easy.

The first step down the road to successful biological farming is balancing the soil minerals. That is the easiest part to address, and though there are budget considerations, you can always take small steps each year towards fixing the soil. Address shortages with plant- and earth-friendly soil correctives, and supply a balanced crop fertilizer that fits the soil and the crop being grown. There is no magic bullet and no one perfect solution, just the right thing for the job at that moment and in that situation. But when it works, it can seem like a miracle.

The second step is supplying some kind of plant-available calcium source. This is especially true for us here in the Midwestern U.S., and mostly true for a lot of other places in the world. There may be crops and soils that do not need added calcium, but if farmers are not supplying calcium, there had better be an abundant, exchangeable amount present in the soil.

Once the first two steps are covered, move on to the rest: soil biology, carbon, nutrient exchange, plant protection and soil health.

The fun in farming and consulting is not in finding a perfect formula that grows perfect crops. It is not that simple. Conventional agriculture has focused on easy fixes using soluble chemistry, mostly NPK, and plant protection through the use of biotechnology and chemicals. This has made many producers into slaves to agribusiness and technology, always passing on both the responsibility, and the profit, to someone else. No wonder our youth want out of farming!

The future of farming will be biological, and the opportunities are huge.

As humans, we have a tremendous job facing us. The farming methods we have been using got us this far, but along the way our environment and health have taken a hit. In hindsight, we can see the problems we have created in the name of cheap food. The future must be, and will be, different. It is now time for "brains, balance and biology." We have a lot of people to feed, and only one Earth to take care of. We need to help the Earth to take care of us, as well.

If "sustainable" is just keeping things going and not falling backwards, it is not good enough. We can do so much more, but to "push" production with tools like commercial nitrogen for crops and confinement and heavy grain diets for cattle may produce a lot of food, but at what cost?

Our other option is to "lead" production. Do everything you can to get the soil healthy and mineralized, and you can't stop the yields. We do not need another standard or more rules and regulations. There are two important questions, both of which need a plan and much thought to be answered: 1) What are you doing to get your soils healthy and mineralized? and 2) What are you doing to get your livestock healthy and comfortable?

At the end of the day, what did I gain, and what did I lose? Any farming system that we talk about will have a carbon "footprint." Biological farming involves doing more tillage, more planting, and using more seeds than conventional farming. But does this add to or reduce the carbon footprint? Which produces more atmospheric carbon: weed control with cultivation or weed control through chemicals? Producing herbicides in a factory requires carbon. The herbicide then gets trucked to the distributors, dealers and farmers — this takes more carbon. After soil application, it changes plant diversity and wipes out soil food, and now I will need even more plant protection. These insecticides and fungicides also need to be manufactured, transported and applied. Chemical-based systems have fewer types of plants feeding their soil life, which means I now need to apply more nitrogen. The production of commercial nitrogen sources uses a whole lot of fuel, which makes it a real carbon user, a destroyer.

Compare this to mechanical weed control or maybe a combination of very minimal chemical weed control and some mechanical. For us on our farm, growing corn involves incorporating green manures and residues, rotary hoeing and cultivating. This does use carbon in the form of tractor fuel for the number of trips across the field, but no one had to haul anything to my farm, and it is even possible for me to grow my own fuel from biofuel-producing crops. Doing mechanical weed control also puts a diversified green manure crop in the form of weeds back into the soil.

Then there is nitrogen. The cheapest, least energy intensive item I can haul to my farm is seeds, grown, gathered and delivered. A truckload of commercial nitrogen or a truckload of seeds for soil building: which takes the most carbon to show up on my farm?

My biological farming system requires more tillage (shallow incorporating cover crops and residues, and several rounds of cultivation), certainly more management, and a whole new set of knowledge. But in the end, the amount of carbon released into the atmosphere compared to the amount sequestered in the soil is a lot better with my farming method compared to the conventional farmer down the road.

If you could walk in my shoes for a year I am sure you would be as optimistic as I am, and as driven as I am to get the word out about the tremendous benefits of biological farming. I am convinced it works because I have seen, tested and tasted the success of this truly sustainable farming system. The opportunities are endless; the project is valuable and achievable.

We have a huge responsibility. The health, wealth and quality of our world is in our hands. We can do better!

Fifty or a hundred years from now, people will look back at this period of time and say, "Oh my gosh, if only they would have known what they were doing, and what was really possible. It was so clean and healthy when they started, and look where conventional agriculture got them."

Bad times make good farmers, researchers and agribusiness people. Maybe things have to get bad enough before they can get better — more pollution, more sick people, more damaged soils and more sick livestock. I truly believe we will need another production system in place once we recognize that the one we are following is not viable for the long term.

I want to go beyond just keeping agriculture the same. What is achievable is way beyond what we are getting today. We have a lot of people to keep fed and healthy, and we need to do it in an environmentally friendly way.

Looking Deeper

I hope this book has challenged you to look deeper. If you have not already read *The Biological Farmer*, it covers a lot of the basics.

Some books that I have really enjoyed reading and studying from include the late Don Schriefer's books, which are available though Acres U.S.A. There are many, many other useful publications available from the

Acres U.S.A. bookstore, but with so many options, the question is: how do you not get confused or sidetracked?

When reading or listening, try to put the information into a category. If we are talking about green manures, cover crops and compost, know where that fits in. If the book is on mineral balance, try to gain a better understanding of how that method or approach can work on your farm. Farmers have much to deal with and it helps to have a plan in place that outlines all aspects of what you want to do before you consider your budget and limitations. Start with the chemistry, the minerals. Then dig in your soils, look at soil structure and soil life. Create that ideal home for soil life and give it time.

So you have just read my story and hopefully through it you have gained a better understanding of this "mineralized, balanced agriculture" approach to farming.

I will end with two questions that we have worked hard to answer on our Otter Creek family farm: What are you doing to get your soils healthy and mineralized? What are you doing to get your livestock healthy and comfortable? When you have a plan in place to answer those questions, you will be well on your way to being a successful biological farmer.

Here's what we do to answer these two questions at Otter Creek Farm:

Question 1

What are you doing to get and keep soils healthy and mineralized?

- We always start with a soil test.
- We then add any mineral that is in short supply, starting with calcium and phosphorus, and use natural mined minerals where possible.
- We provide a balanced diet for the crop, including humates, kelp, plant-available calcium, naturally mined minerals and all the trace elements.
- Yard manure is applied to fields that will be planted to corn.
- We make compost (stacked, aged manure), and spread it on hay fields during the summer. We use a slinger spreader to provide a light coat on a lot of acres.

- We apply calcium and other soil correctives in the fall, a crop fertilizer in early spring when planting, and compost or aged manure in the summer on forage crops.
- We have a tight crop rotation and grow a large diversity of plants to extract natural soil minerals and feed soil life. We also take every opportunity to grow green manure crops, including buckwheat, grasses, fall rye, early spring oats, peas and brassicas.
- We till only when necessary, using either shallow incorporation and/or subsoiling for air and water management.
- We shallow incorporate green manure crops and residues to protect the soil and feed soil life. We also like to keep a blanket of residues on the soil.
- To promote soil biology, we grow a large diversity of plants, use minimal tillage, apply compost, and inoculate seeds and compost with live organisms.
- For our crop rotation, on the flat fields we like to use a 2 year hay/1 year corn rotation. Sometimes a second year of corn is added, especially if corn silage is harvested followed by a rye cover crop. Hilly or stony ground is left longer in forages.
- Soybeans and peas are added to the rotation when and where they fit.

Question 2

What are you doing to get livestock healthy and comfortable?

- We always strive to grow high-quality feeds.
- We balance the ration and provide all the extras, including kelp, direct-fed microbials, yeast, vitamins, Dynamin and CharCal.
- We always feed a high-forage diet, with as much pasture as possible.
- We raise our calves on whole milk with added nutritional products, and do not wean until the calves are at least 10 weeks old.
- All cattle have access to the outdoors and fresh air year round, and are on bedding packs in the winter.

And lastly, what we *do not* do is as important as what we do.

Bibliography

Chapter 8

Barak, P., B.O. Jobe, A. Krueger, L.A. Peterson, and D.A. Laird. 1997. *Effects of long-term soil acidification due to agricultural inputs in Wisconsin.* Plant and Soil 197:61-69.

Brady, Nyle C. and Ray R. Weil. 2997. *The Nature and Property of Soils.* Pearson Prentice Hall, 965 pp.

Mengel and Barber, 1974. *Role of Nutrient Uptake per Unit of Corn Root Under Field Conditions.* Agronomy Journal 66: 399-402.

Roots, Growth and Nutrient Uptake, Purdue University Dept. of Agronomy Publication #AGRY-95-08

Schriefer, Donald L. 2000. *From the Soil Up.* Acres U.S.A. 274 pp.

Chapter 9

Barak, P., B.O. Jobe, A. Krueger, L.A. Peterson, and D.A. Laird. 1997. *Effects of long-term soil acidification due to agricultural inputs in Wisconsin.* Plant and Soil 197:61-69.

Beltsas, John. Total Growers Services Pty Ltd. reported the results of his experiment adding calcium and boron to banana plants in a letter Gary received. "After 10 years of unsuccessfully trying to control this disease with ag-chem, I am pleased to say that we have witnessed twelve months of control," Beltsas wrote.

Datnoff, Lawrence E., Wade H. Elmer and Don M. Huber, editors. 2007. *Mineral Nutrition and Plant Disease.* The American Phytopathological Society. 278 pp.

FitzGerald S., B. Goldman, A.A. Dickinson and S.D. White, 2003. *A survey of Sigatoka leaf disease (Mycosphaerella musicola Leach) of banana and soil calcium levels in North Queensland.* Australian Journal of Experimental Agriculture, 43 (9): 1157 – 1161.

Norton, Darrell. 1997. *Stopping Erosion with Gypsum and PAM*. United States Department of Agriculture Research Bulletin.

Palta, Jiwan P. *Impact of Calcium Nutrition on Tuber Quality and Yield.* 1997. The Badger Commentator, August issue, pp 25-42.

Phelan, P. L. 2004. Connecting Belowground and Aboveground Food Webs: The role of organic matter in biological buffering, Chapter 5, in *Soil Organic Matter in Sustainable Agriculture*, F. Magdoff and R. R. Weil, eds. CRC Press, Boca Raton, FL.

Chapter 11

Brady, Nyle C. and Ray R. Weil. 2007. *The Nature and Property of Soils.* Pearson Prentice Hall, 965 pp.

Michigan Field Crop Ecology. Michigan State University Extension Bulletin E-2646. Published December, 2000.

Chapter 12

Snapp, Sieg. 2009. *Is it possible to build soil organic matter while simultaneously decomposing soil organic matter to provide nitrogen?* Organic Broadcaster Jan/Feb 2009. Published by The Midwest Organic and Sustainable Education Service.

Brady, Nyle C. and Ray R. Weil. 2007. *The Nature and Property of Soils.* Pearson Prentice Hall, 965 pp.

Magdoff, Fred and Harold van Es, 2000. *Building Soils for Better Crops.* Published by SARE Outreach. 230 pp.

Chapter 13

Managing Cover Crops Profitably. 2007. Handbook Series Book 9, Published by the Sustainable Agriculture Network, Beltsville, MD. 244 pp.

Chapter 14

Schriefer, Donald L. 2000. *From the Soil Up.* Acres U.S.A. 274 pp.

Schriefer, Donald L. 2000. *Agriculture in Transition.* Acres U.S.A. 240 pp.

Index